Packaging for Nonthermal Processing of Food

The *IFT Press* series reflects the mission of the Institute of Food Technologists—advancing the science and technology of food through the exchange of knowledge. Developed in partnership with Blackwell Publishing, *IFT Press* books serve as leading-edge handbooks for industrial application and reference and as essential texts for academic programs. Crafted through rigorous peer review and meticulous research, *IFT Press* publications represent the latest, most significant resources available to food scientists and related agriculture professionals worldwide.

Packaging for Nonthermal Processing of Food

EDITED BY

Jung H. Han, Ph.D.

Department of Food Science

University of Manitoba

Winnipeg, Manitoba

Dr. Jung H. Han is an assistant professor of dairy/food processing and packaging at the University of Manitoba, Canada. He is a professional member of the Institute of Food Technologists (IFT). He is an associate editor of *Journal of Food Science* and an editorial board member of *Food Research International*. Dr. Han has served as secretary, chair-elect, and chair of the Food Packaging Division of IFT. He is editor of *Innovations in Food Packaging*.

Blackwell Publishing Professional
2121 State Avenue, Ames, Iowa 50014, USA

Orders: 1-800-862-6657
Office: 1-515-292-0140
Fax: 1-515-292-3348
Web site: www.blackwellprofessional.com

Blackwell Publishing Ltd
9600 Garsington Road, Oxford OX4 2DQ, UK
Tel.: +44 (0)1865 776868

Blackwell Publishing Asia
550 Swanston Street, Carlton, Victoria 3053, Australia
Tel.: +61 (0)3 8359 1011

Authorization to photocopy items for internal or personal use, or the internal or personal use of specific clients, is granted by Blackwell Publishing, provided that the base fee is paid directly to the Copyright Clearance Center, 222 Rosewood Drive, Danvers, MA 01923. For those organizations that have been granted a photocopy license by CCC, a separate system of payments has been arranged. The fee codes for users of the Transactional Reporting Service are ISBN-13: 978-0-8138-1944-0/2007.

Cover image: E-beam unit for packaged frozen meats (*left*) and pulsed UV/light emission for liquid foods (*right*).

First edition, 2007

Library of Congress Cataloging-in-Publication Data

Packaging for nonthermal processing of food / edited by Jung H. Han.—1st ed.
      p. cm.—(IFT Press series)
   Includes bibliographical references and index.
   ISBN-13: 978-0-8138-1944-0 (alk. paper)
   ISBN-10: 0-8138-1944-X (alk. paper)
  1. Food—Packaging. I. Han, Jung H.

   TP374.P328 2007
   664′.09—dc22
                                                    2006036529

The last digit is the print number: 9 8 7 6 5 4 3 2 1

Titles in the *IFT Press* series

CONTENTS

CONTRIBUTORS

Aaron L. Brody
Packaging/Brody, Inc., P.O. Box 956187, Duluth, GA 30096, USA

Jung H. Han
Department of Food Science, University of Manitoba, Winnipeg,
Manitoba, Canada R3T2N2

A. R. de Jong
TNO Quality of Life, Utrechtseweg 48, P.O. Box 360, Zeist,
The Netherlands

Arnold W. Hydamaka
Department of Food Science, University of Manitoba, Winnipeg,
Manitoba, Canada R3T2N2

Joan C. Junkus
Department of Finance, DePaul University, 1 East Jackson Boulevard,
Chicago, IL 60604, USA

Vanee Komolprasert
Office of Food Additive Safety, Center for Food Safety and Applied
Nutrition, US Food and Drug Administration, 5100 Paint Branch
Parkway, College Park, MD 20740, USA

John M. Krochta
Department of Food Science and Technology, University of California,
Davis, One Shields Avenue, Davis, CA 95616, USA

Seacheol Min
Department of Food Science and Technology, University of California,
Davis, One Shields Avenue, Davis, CA 95616, USA

M. A. H. Rijk
TNO Quality of Life, Utrechtseweg 48, P.O. Box 360, Zeist,
The Netherlands

Kevin C. Spencer
Spencer Consulting, 424 Selborne Road, Riverside, IL 60546, USA

W. D. van Dongen
TNO Quality of Life, Utrechtseweg 48, P.O. Box 360, Zeist,
The Netherlands

James T. C. Yuan
Food Safety and Healthcare Applications, American Air Liquide, Inc.,
5230 S. East Avenue, Countryside, IL 60525, USA

Q. Howard Zhang
Food Safety Intervention Technologies Research Unit, United State
Department of Agriculture, Eastern Regional Research Center,
600 East Mermaid Lane, Wyndmoor, PA 19038, USA

Yicheng Zong
Department of Food Science, University of Manitoba, Winnipeg,
Manitoba, Canada R3T2N2

PREFACE

During 2005 IFT (Institute of Food Technologists) Annual Meeting (July 15–20, 2005, New Orleans, Louisiana, USA), Food Packaging Division and Nonthermal Processing Division co-sponsored two symposia under the titles of "Advances in Packaging Technology Required for Implementation on Novel Food Processes" and "Active Packaging for Nonthermal Processing." I was privileged to host these symposia as a Chair-elect of Food Packaging Division with Dr. James Yuan who was the Chair of Nonthermal Processing Division. Both symposia were successful as they provided valuable opportunities for general participants and invited speakers to share their scientific interests and practical experiences. I recognized instantly that the papers of these symposia were valuable and should be published to provide current information in a stable and concrete form. I thank IFT for accepting my proposal to contribute a book to the IFT Press Series.

Nonthermally processed food products have unique quality parameters compared to conventional food products processed by thermal treatment. Some of these new quality parameters might not be essential for the thermal processes because they were new parameters which used not to be considered for conventional thermal processes. Therefore, nonthermal processes have new requirements of processing and packaging to protect the quality of nonthermally processed foods. The conventional packaging design and materials should be changed accordingly for nonthermally processed foods. Critical protective barrier properties of packaging materials must remain to prevent chemical, physical, or microbial degradation of the contents after nonthermal processing. This book discusses the need to understand the details of process, product, and packaging material interactively for the selection of commercial products that are ready for extended shelf life and consumer testing. In addition, the critical role of information carried by packaging materials is discussed to make a new product produced by a novel process attractive to consumers.

This book could not be possible without the contribution of the chapter authors. My special thanks goes to IFT staff and Blackwell Publishing staff for initiating and completing the publication processes.

Jung H. Han, Ph.D.

Packaging for Nonthermal Processing of Food

Chapter 1

PACKAGING FOR NONTHERMALLY PROCESSED FOODS

Jung H. Han

Introduction

Since nonthermally processed food products have unique quality parameters when compared with food products processed by conventional thermal treatment, the packaging design and materials used for the nonthermally processed food products should be changed accordingly. Advances in packaging technology are required for implementation in novel food processes. A number of novel thermal and nonthermal processing methods are actively undergoing research and development in industrial, academic, and government laboratories. A key step that now needs addressing is finding the best packaging materials for commodities processed by nonthermal procedures such as high pressure, pulsed electric fields, ultraviolet (UV), irradiation, microfiltration, active packaging (oxygen scavenging or antimicrobial packaging), or biopreservation (antagonistic culture), which preserve the benefits of improved product quality imparted by these emerging preservation technologies. Critical protective barrier properties of packaging materials must be preserved to prevent chemical, physical, or microbial degradation of contents after nonthermal processing. This book discusses the need to understand details of process, product, and packaging material interactively for selecting commercial products that are ready for extended shelf life and consumer testing. In addition, the critical role of information carried by packaging materials in making a new product produced by a novel process attractive to consumers is discussed.

3

Nonthermal Processing of Foods

Food processors traditionally utilized thermal processes, that is, cooking, blanching, pasteurization, and sterilization, to inactivate microorganisms, enzymes, and other chemical reactions in food materials as well as to cook raw foods for extending the period of desirable quality and safety level. Because of numerous practical applications of heat treatments with various types of foods, from prehistoric age until today, many chemical and physical changes taking place in foods after the thermal process have been understood. Not only the changes in the nature of food products after thermal processes, but also the chemical interactions between thermally processed foods and common food packaging materials are well identified.

Characteristics of Nonthermal Processes

Nonthermal processes are food preservation methods to inactivate spoilage and pathogenic microorganisms at temperatures below those used for thermal pasteurization without significant changes to flavor, color, taste, nutrients, and functionalities (Min and Zhang 2005). They involve the use of high electric power, high pressure, high intensity of light/radiation, microfilters, chemicals, or antagonistic cultures. Electric power, pressure, light emission, radiation dose, microsieves, and/or natural chemicals have not been used traditionally for food preservation. Compared with heat treatment, these nonthermal treatments are less studied for their effects on chemical, physical, and microbiological changes in food products. The quality attributes of nonthermally treated foods are obviously different from those of thermally processed food products. Moreover, the chemical reactions involved in food quality deterioration following nonthermal treatment would be different from the reactions in heat-treated food products (Figures 1.1–1.3).

Among many alternative nonthermal pasteurization treatments, pulsed electric field and high-pressure processing are the most investigated treatments despite their short history compared with other treatments such as irradiation and chemical treatments (Butz and Tauscher 2002). Both technologies allow the inactivation of vegetable microorganisms but fail to destroy spores when they are applied alone (Devlieghere, Vermeiren, and Debevere 2004). Because of these relatively

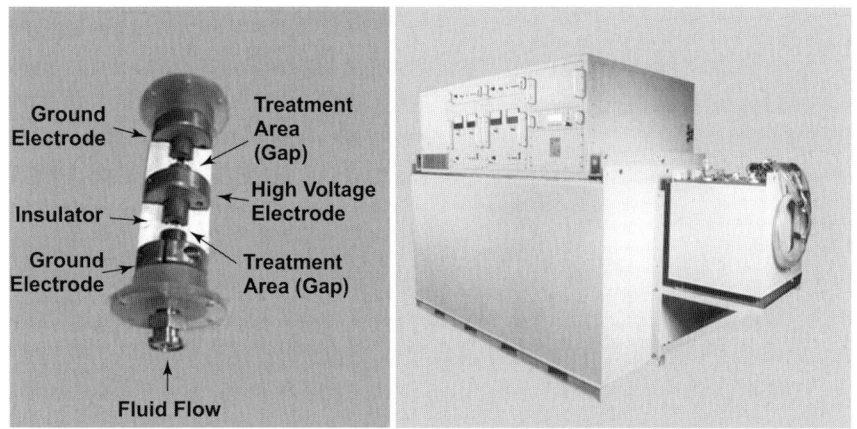

Figure 1.1. Pulsed electric field chamber structure and exterior of processing unit (courtesy of Diversified Technologies, Inc., and Ohio State University).

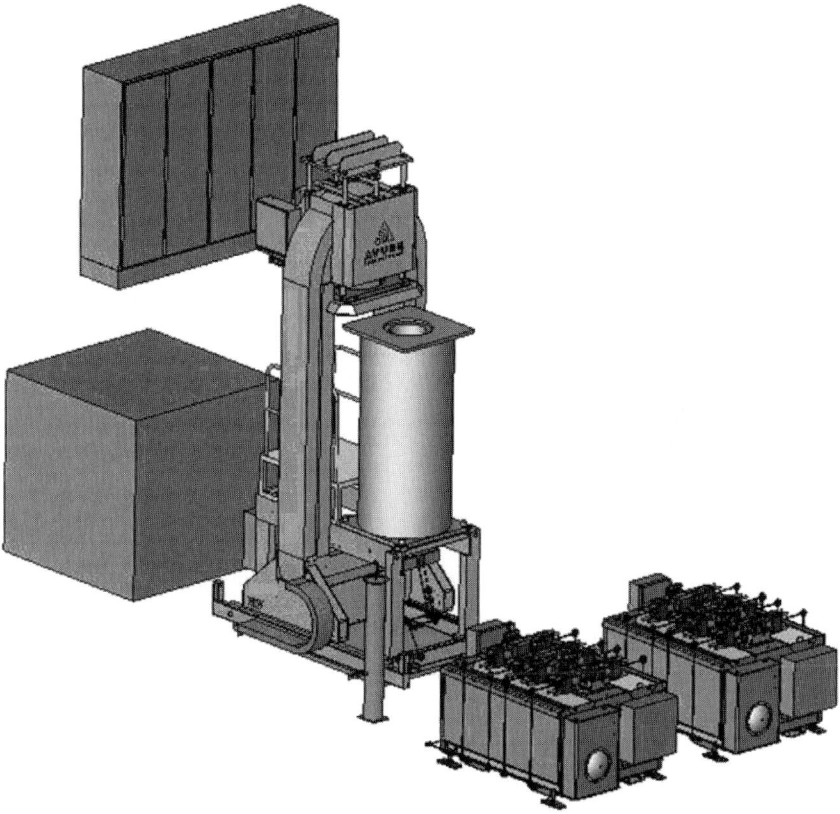

Figure 1.2. High-pressure processing system. Food will be inserted into the cylindrical pressure vessel (courtesy of Avure Technologies, Inc.).

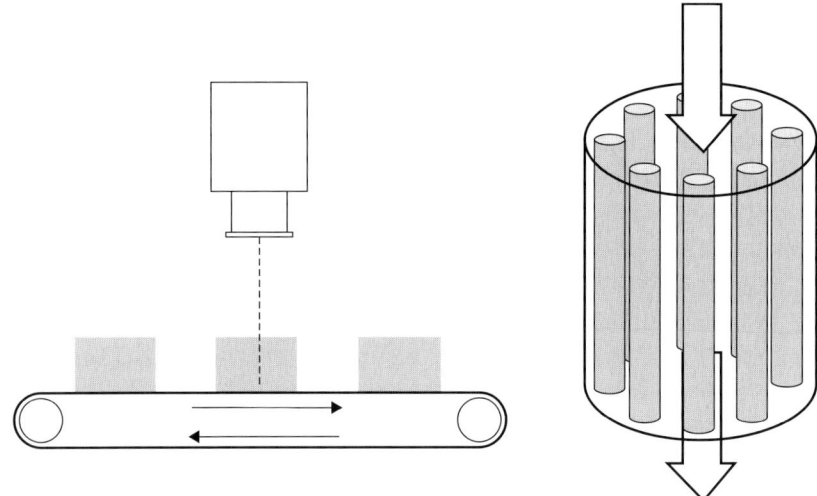

Figure 1.3.   e-beam unit for packaged frozen meats (left) and pulsed UV/light emission for liquid foods (right). e-beam radiates on the packaged frozen case meats, and light illuminates from light tubes to liquid foods inside the chamber.

mild conditions of nonthermal processes compared with heat pasteurization, consumers may be easily satisfied by the more fresh-like characteristics, minimized degradation of nutrients, and the perception of high quality (Butz and Tauscher 2002; Min and Zhang 2005) (Table 1.1).

Concerns of Nonthermal Processing

Bacillus stearothermophilus has been used as an index microorganism for the standard evaluation of thermal processes. For other specific foods with extreme conditions of pH, water activity, and solute concentration, other spore-forming bacteria have been used to verify adequate heat treatment. There are many but standardized data tables of the values of D (time) and Z (temperature) for these standard microorganisms, and the effect of these thermal treatments is evaluated based on their F value. However, owing to the diverged resistance of microorganisms to various nonthermal processes such as electric field, pressure, irradiation, or chemicals, it is impossible to identify the appropriate standard microorganisms and their quantitative resistance index to the nonthermal processes. Various scientific studies have been

Table 1.1. New nonthermal food-processing methods (reprinted from Butz and Tauscher 2002, p. 280, with permission from Elsevier Ltd.).

Process	Description	Critical Factors	Mechanism of Inactivation	Status
Ultraviolet (UV) light/pulsed light	UV radiant exposure, at least 400 J/m^2 Intense and short-duration pulses of broad spectrum (ultraviolet to the near-infrared region)	Transmissivity of the product, the geometry, the power, wavelength, and arrangement of light source(s), and the product flow profile	DNA mutations induced by DNA absorption of the UV light	Used for disinfection of water supplies and food contact surfaces
Ultrasound	Energy generated by sound waves of 20,000 Hz or more	The heterogeneous and protective nature of food (e.g., inclusion of particulates) severely curtails the singular use of ultrasound for preservation	Intracellular cavitation (micromechanical shocks that disrupt cellular, structural, and functional components—cell lysis)	Combination with, for example, heat and pressure has certain potential
Oscillating magnetic field (OMF)	Subjecting food sealed in plastic bags to 1–100 OMF pulses (5–500 kHz, 0–50°C, 25–100 ms)	Consistent results concerning the efficacy of this method are needed	Controversial results on the effects of magnetic fields on microbial populations	Application at the moment not considered

(continued)

Table 1.1. New nonthermal food-processing methods (reprinted from Butz and Tauscher 2002, p. 280, with permission from Elsevier Ltd.). (*Continued*)

Process	Description	Critical Factors	Mechanism of Inactivation	Status
Pulsed electric field (PEF)	High-voltage pulses to foods between two electrodes (<1 s; 20–80 kV/cm; exponentially decaying, square wave, bipolar, or oscillatory pulses at ambient, subambient, or above-ambient temperatures)	Electric field intensity, pulse width, treatment time, temperature, pulse wave shapes, type, concentration, and growth stage of microorganism, pH, antimicrobials, conductivity, and medium ionic strength	Most theories studied are electrical breakdown and electroporation	Different laboratory- and pilot-scale treatment chambers designed and used for foods; only two industrial-scale PEF systems available
High-pressure processing	Liquid/solid foods, with or without packaging (100–800 MPa, <0°C to >100°C, from a few seconds to >20 min instantaneously and uniformly throughout a mass of food independent of size, shape, and food composition)	Pressure, time at pressure, temperature (including adiabatic heating), pH, and composition	Denaturation of enzymes and proteins, breakdown of biological membranes, and cellular mass transfer affected	In use since 1990 (Japan, United States, France, and Spain); current pressure processes include batch and semicontinuous systems

8

conducted to determine the resistance of significant microorganisms to nonthermally processed foods. For example, there are many studies by civil and environmental engineers on the resistance of *Escherichia coli* or other bacteria to UV radiation treatments in drinking water. These researches dealt with selected microorganisms and determined their resistance (*D* and *Z* values for time and intensity, respectively) to UV light. However, they did not suggest standard target microorganisms for UV light treatment, like the *B. stearothermophilus* for heat treatment, because their projects were oriented to inactivating coliform bacteria and parasites.

Nonthermal processes can be applied as a part of combined processes with other nonthermal processes, heat treatments, or chemical treatments. In these combinations of multiple processes, it is harder to select the target microorganism as a standard because the other combined processes affect the resistance of the target microorganisms to the nonthermal processes. However, the effects of these nonthermal processes on microbial inactivation would be increased synergistically with other treatments by maintaining fresh-like taste and retaining color and nutrients of foods.

Besides these technical aspects of nonthermal processes, there are more significant nonscientific factors such as consumers' acceptance of the new technologies, which also govern the commercialization of nonthermal processes strongly. Since food is a critical consumer product, applying any new technology that consumers are not familiar with, regardless of the positive results of scientific risk assessment, is a very sensitive issue.

Official permission from governmental regulatory agencies to use the new technology is one of the requirements for the commercialization of new nonthermal processes, which can also be considered as a partially nonscientific parameter. Since food-related legislation process is very political, the regulatory issues of the permission to use any nonthermal process are also political. Food industry, regulatory agencies, consumer groups, and nonthermal equipment industries may have different voices to the same issue. Furthermore, there are not enough scientific data and results to provide one solution to these diverged parties of interests. Therefore, it is very important to identify the current development status of technology commercialization and for scientific societies and academic institutes to provide a fair milestone to these interest groups.

Nonthermal food processes can substitute conventional heat treatment, at least partially, depending on the nature of commodities or foods. The use of nonthermal processes for food preservation will become more popular in the near future. However, the complexity of scientific and nonscientific parameters could influence the successful utilization of these nonthermal processes. This complexity can be simplified by tight communication and collaboration between various interest groups.

Food Packaging

Roles of Food Packaging Systems

Food is packaged for storage, preservation, and protection traditionally for a long time. These three are the basic functions of food packaging that are still required today for better maintenance of quality and handling of foods. In addition to these primary functions of food packaging, more superficial functions are required for food marketing, distribution, and consumer-related issues, which are to provide required information, handling and dispensing convenience, sales promotion, and stock management. No matter what new fancy function of packaging is explored, the first priority should be serving the basic functions of food packaging.

Generally, the main goal of food packaging is to maintain the quality of packaged foods during distribution. However, following the evolution of modern society and lifestyle, the significance of several functions of packaging is also shifting from one aspect to another. There is a trend of changes in the priority of these functions with time and social circumstances. The distribution of priority of each packaging function is highly dependent on the commodities and, therefore, the properties of packaged foods. Owing to cost-saving high-quality packaging materials, the basic functions of packaging materials and systems are not considered as the most significant functions. Instead, the importance of superficial functions becomes emphasized when a new food package is designed. The major functions of food packaging have moved from containment, preservation, and protection to convenience and sales promotion, which indicates that the role of food packaging is shifting from technical functions to socioeconomic functions.

After the tragedy of September 11, 2001, a new critical function appears to be the top priority of food packaging purpose: safety and security. More functions that could secure the integrity of packaging are considered as very important roles of packaging, including tamper-evidence packaging, antimicrobial packaging, freshness indicator, time-temperature integrator, and electronic coding system.

The fast development of nonthermal processing systems is another factor that affects the changing trends in the functions of food packaging. Most food packaging materials and systems have been designed for fresh produce, fresh meats, or heat-processed (i.e., cooked) foods. The commercialization of a newly developed nonthermal processing system essentially requires deep research on the packaging material properties and the interactions between the packaging materials and nonthermally processed foods. The packaging for nonthermally processed foods may necessitate extra functions of packaging for successful commercialization.

Packaging for Nonthermal Processing

Food products and packaging materials interact with each other. These interactions include the migration of packaging ingredients into food products, absorption of hydrophobic food ingredients such as flavors and colorants into packaging structures, and the oxidative reactions of residual sanitizers used for packaging surface disinfection with contacted food surfaces.

Conventional food manufacturing processes use high temperature for pasteurization just before or after the food products are packaged. Special considerations of the effect of high temperature on the physical and chemical properties of packaging materials are essentially required. Packaging material should have strong resistance to the thermal effect. This resistance may be determined by phase-transition temperature, elastic modulus, physical strength, and barrier properties with response to the temperature of the thermal process. For thermal processing, the heat resistance of various microorganisms has been determined, and the effects of thermal process on the survival of the microorganisms have been evaluated by thermal death time, F value. Therefore, the changes in packaging material characteristics because of heat should be minimized with this F value condition. Thermal

processing of packaged foods may change the original physical characteristics of packaging systems, such as surface adhesion, sealability
of packaging, and partial pressure changes due to the gas permeability
of packaging materials and condensation of moisture inside packaging.
All of these chemical, physical, and biological changes of foods
and packages after thermal processes are currently predictable and
preventable.

Nonthermal processes can be operated with aseptic processing units.
For pumpable foods, nonthermal processing units can replace heat-
exchanger units of aseptic processing systems. To maintain the sterility
of nonthermally processed foods, the processed foods must be packaged before the sterile foods are exposed to atmosphere. Therefore,
aseptic processing units are required after the nonthermal processes
such as pulsed electric field, high-pressure process, irradiation, pulsed
light emission, microfiltration, or inoculation of antagonistic bacteria.
These processes generally induce damage to microorganisms, such as
irreversible membrane perforation or deformation, thus inactivating
them. These treatments also can cause serious damages to the cells and
tissues of animal products and are unsuitable for meat or fish (Lelieveld
2006). These damages to vegetables may be desirable in most cases
because of the increase in the bioavailability of nutrients after treatment
(Lelieveld 2006).

To maintain the sterility of nonpumpable foods, the foods should
be packaged before nonthermal processes such as pulsed light
emission, irradiation, batch high-pressure process, or antimicrobial
packaging. Therefore, similar to the heat resistance of packaging
materials to thermal treatment, the packaging materials for nonthermal
processing should have appropriate resistance to high-energy light,
irradiation, pressure, or chemicals. This means that the chemical
and physical requirements of packaging materials for nonthermal
processing are different from those for thermally processed foods.
Therefore, to identify these special requirements of packaging materials for nonthermal processing, it is necessary to understand the
process parameters and microbiocidal mechanisms/kinetics of the nonthermal process and their effects on mechanical and physical properties
of packaging materials.

Besides the mechanical and physical characteristics of packaging
materials, various other factors of food packaging systems should be

considered in designing the package for the nonthermal processes, which may include, as in, for example, high-pressure process, volume of the package, headspace gas, dissolved oxygen in foods, and deformation characteristics of packaged foods (Balasubramaniam *et al.* 2004).

New Requirements of Food Packaging Systems for Nonthermal Processing

For the commercialization of nonthermal processing in food industry, many scientific and nonscientific parameters of packaging systems should be studied in addition to research and development of machinery and equipment. These parameters should include packaging requirements, material characteristics, consumers' acceptance, regulations. Balanced research and development in these areas will facilitate the successful commercial use of nonthermal processes for preservation.

Characteristics of Packaging Materials

Packaging materials for nonthermal processing should have strong resistance (physical and mechanical properties) to the nonthermal process mechanisms. For example, packaging materials for high-pressure processing should be restored from the deformation under pressure to their original shape without any mechanical and physical change of properties. The packaging materials for irradiation should be chemically stable under the radiation dose without depolymerization or significant changes in elastic modulus of the packaging materials. For pulsed UV/white-light emission process, the packaging material must be transparent during pulsed light emission. There is no general requirement for packaging materials for all nonthermal processes. However, from the above examples, most characteristics of packaging materials required for the various nonthermal processes are related to the barrier properties of the packaging materials. This is due to the satisfaction of the primary functions of packaging systems: containment, protection, and preservation.

Nonthermal processes do not utilize increased temperature to inactivate decomposing microorganisms and enzymes. This is the biggest advantage of nonthermal processes because this low-temperature pasteurization does not overcook food and/or degrade foods thermally. Furthermore, this low-temperature treatment also widens the selection of packaging materials and systems. Owing to the low-temperature treatment, the packaging system does not require high melting temperature for heat seal. Low-temperature sealing methods can utilize various polymers and sealants, if required, or cold sealing using adhesives. These methods produce far less volatile odor of plastics, additives, and printing solvent. This is very beneficial to high-fat foods and frozen/refrigerated foods.

Consumer Acceptance and Regulation

Food is one of the consumer products. Consumers purchase packaged foods. Therefore, packages are the interface linking food products and consumers. Consumers react to a package that represents internal food product with their own favorites. For the commercial use of nonthermal processes that consumers are not familiar with, food packaging materials should provide information on the new processes to reduce the reluctance among consumers. It is better to maintain conventional design and structure of the package for the new products and technologies. The same or similar packaging design would minimize the reluctance among consumers due to the fear of new design or new processing. For the extended commercialization of nonthermal processes, it is necessary to study consumers' acceptance of the nonthermally processed packaged foods.

New regulation to control the use of nonthermal processes should be established fairly based on scientific evidence. The regulation is the most critical factor for the further use of nonthermal processes. It can facilitate or limit the use of nonthermal processes. This regulation should include specific guidelines on the operation conditions of nonthermal processes as well as the approval of packaging materials and required labeling constituents that may not threaten consumers.

Table 1.2 summarizes the opportunities and drawbacks of these nonthermal technologies. Like various characteristics of the nature of foods, there are a variety of characteristics of these nonthermal technologies. Marriage of these characteristics of food products and processes will provide for very favorable and promising commercialization of the new nonthermal technologies.

Table 1.2. Important opportunities and drawbacks of new nonthermal techniques (Reprinted from Devlieghere *et al.* 2004, p. 282, with permission from Elsevier Ltd.).

New Preservation Techniques	Opportunities	Drawbacks
High hydrostatic pressure	High retention of nutrients and vitamins	Discontinuous for solid, viscous, and particulate foods Semicontinuous for liquid foods
	High fresh-like organoleptic quality	High investment costs
	Applicable for acid foods (as spores will not germinate in acid foods)	Spores are not sensitive
	Spores can be inactivated when combined with heat or lactoperoxidase system or lysozyme	Bacteria have been shown to become resistant
Pulsed electric fields	Continuous process is possible	Upscaling of equipment is still under development
	High retention of nutrients and vitamins	Limited to liquid products
	High fresh-like organoleptic quality	Spores are not sensitive
	Applicable for acid foods (as spores will not germinate in acid foods) Effect of combinations with other preservation methods to inactivate spores is under investigation	Effectivity depends on the electrical conductivity of the food
Active packaging	Oxygen scavengers in sachets are effective	Oxygen scavengers incorporated in a packaging film often show effectivity limited Incompatibility with legislation
	Enables a surface treatment of food products	Amount of active compound migrating is often not substantial

(continued)

Table 1.2.   Important opportunities and drawbacks of new nonthermal techniques (Reprinted from Devlieghere *et al.* 2004, p. 282, with permission from Elsevier Ltd.). (*Continued*)

New Preservation Techniques	Opportunities	Drawbacks
	Facilitates processing	Active compounds have to be thermostable when incorporated in plastic films
Natural antimicrobial compounds	Green labeling	Often expensive Interaction with food ingredients Low water solubility
	Natural image	Change the organoleptic properties
Bacteriocins	Natural image	Narrow activity spectrum Spontaneous loss of bacteriocinogenicity Limited diffusion in solid matrixes Inactivation through proteolytic enzymes Interaction with food ingredients Bacteriocin-resistant bacteria
Protective cultures	Green labeling	Sometimes difficult to apply Heat unstable
	Natural image	Effectivity is not always proven in food products

References

Balasubramaniam, V.M., Ting, E.Y., Stewart, C.M., and Robbins, J.A. 2004. Recommended laboratory practices for conducting high-pressure microbial inactivation experiments. *Innovative Food Science & Emerging Technologies*. 5: 299–306.

Butz, P., and Tauscher, B. 2002. Emerging technologies: chemical aspects. *Food Research International*. 35(2/3): 279–284.

Devlieghere, F., Vermeiren, L., and Debevere, J. 2004. New preservation technologies: possibilities and limitations. *International Dairy Journal*. 14(4): 273–285.

Lelieveld, H.L.M. 2006. Pulsed electric field pasteurization of foods. *New Food*. 9(1): 31–33.

Min, S., and Zhang, Q.H. 2005. "Packaging for non-thermal food processing." In *Innovations in Food Packaging*, edited by J.H. Han, pp. 482–500. Oxford, UK: Elsevier Academic Press.

Chapter 2

THE ROLE OF ACTIVE PACKAGING IN NONTHERMAL PROCESSING SYSTEMS

Aaron L. Brody

Introduction

This chapter lays foundations for the role of active packaging in the multiple emerging technologies of nonthermal food processing. In this chapter are included definitions for active packaging, issues of active packaging relative to nonthermal processing, definitions for various nonthermal processing technologies, and issues relevant to nonthermal processing. Even more important than the specifics of the individual disciplines are issues at the interface of active packaging and non-thermal processing. This chapter concludes by attempting to forecast the future for active packaging with nonthermal processing.

Definitions

Active Packaging

Conventional mainstream packaging such as paperboard, metal, glass, and plastic may be classified as gas and water vapor barriers or non-barrier, such as paperboard, unless coated.

Active packaging senses changes in the internal or external environment and responds by altering the package properties.

Intelligent Packaging

Active packaging is different from intelligent packaging, which senses and signals. In the future, however, we might expect the conversion of intelligent into active packaging since almost anything that can be measured can usually be controlled (Table 2.1).

Active Packaging: Moisture Control

Probably the largest unit volume application of active packaging is moisture control, largely in nonfood applications. Desiccants such as silica gel in moisture-permeable plastic pouches are common in distribution packages of electronic devices, instruments, medical analytical kits, and so on. The water vapor-permeable plastic may be

Table 2.1.   Active packaging technologies.

- Moisture
  - Remove excess
    - Liquid
      - Purge
      - Absorbents
      - Incorporate into pad
    - Relative humidity
      - Incorporate into package structure
      - Incorporate into cartridge
      - Add to control relative humidity
      - Usually to elevate level
- Light and related radiation blocking
  - To control the amount of reaching contents
- Gas: to control levels
  - Oxygen
  - Carbon dioxide
  - Ethylene
  - Water vapor
  - Odors
- Antimicrobials
- Self-heating
  - Exothermic reaction of water and mineral constituent
- Self-cooling
  - Evaporation of liquid
  - Expansion of compressed gas

spun-bonded polyolefin or even coated porous paper. The desiccant also may be in moisture-barrier plastic cartridges with molded minute openings. In a relatively new development, the desiccant may be adsorbed onto the surfaces of channeled polyolefin sheet developed by CSP Corporation (Auburn, AL).

Another major application of moisture controllers, numbering in billions of units, is in liquid absorbent pads inserted in trays of fresh red meat, poultry, seafood, and fresh produce. Their purpose is to remove the bulk of excess liquid purge, drip, or condensate arising from squeeze or aging of the product plus gravity. The pads may be constituted of wood pulp fibers within perforated polyethylene film or polymers such as carboxymethyl cellulose in cavities topped with spun-bonded polyolefin.

Because purge may represent an excellent microbiological growth medium, it may be enhanced with antimicrobials and even antioxidants (Paper Pak Industries' XtendaPak).

Active Packaging: Oxygen Control

Oxygen is a major deteriorative biochemical and aerobic microbiological growth stimulant vector in food products. Its removal and subsequent exclusion by gas-barrier packaging are key factors in food preservation. The notion of completing oxygen removal started with mechanical methods such as vacuum, steam, inert gas, etc., with reactive mechanisms leading to the concept of oxygen scavengers.

Scavengers, sometimes also called absorbers, remove excess oxygen from the internal package environment. They may be incorporated in sachets (small pouches) or labels affixed to the package or incorporated into package material/structure.

Most commercial oxygen scavengers today are ferrous iron, which requires water for activation to react with oxygen, to form ferric iron, a dark-colored compound. Organic oxygen scavengers include unsaturated hydrocarbons, nylons, and benzyl acrylate copolymers.

Classical antioxidants such as ascorbic acid, butylated hydroxy-anisole/butylated hydroxytoluene (BHA/BHT), and tocopherol have been added to package structures to facilitate their migration from the adjacent plastic into the contained food. A major challenge is to achieve permeation through the product contents to retard oxidation throughout the product.

In some situations, excess (≫20.9% in the package headspace) oxygen may be preferred to maintain red meat color, to retard fresh vegetable aerobic respiration, to obviate respiratory anaerobiosis in respiring produce, to maintain a specific oxygen level, or to retard microbiological anaerobiosis, a questionable effect.

Oxygen may be added to the package interior by the use of high gas transmission package materials/structures: such package structures may function by differential pressure between external air atmosphere and internal low oxygen environment, by highly gas-permeable specialized polymers, by mineral-filled polymers, and by microperforation of the package material. Oxygen passage by permeation is temperature sensitive.

IntelliPack™ is a plastic structure containing side-chain polymers that more readily permeate gas, increasing in permeation rate with increasing temperature.

Oxygen generators such as percarbonates incorporated into the plastic package structure have been employed to elevate oxygen in food packages.

Active Packaging: Carbon Dioxide Control

Carbon dioxide is a known antimicrobial agent whose activity is probably by pH reduction and whose level should be controlled to optimize the desired effect. CO_2 may be added to complement existing procedure/microbiological respiratory CO_2, or to maintain the existing level by generating it from chemical reactions during distribution. In excess, however, CO_2 is able to injure the product quality, and hence sometimes it may be advisable to remove excess from modified atmosphere (MAP) as in packaged respiring produce. On the other hand, at least one processor has intentionally over-carbonated to produce what it regards as a desirable acidic/gassy mouth feel.

Gas control issues include the following: control to specific concentrations in the headspace of the package; ability of the gas or gases to pass throughout the food/beverage mass in equilibrium with the headspace gas; ability to pass sufficient gas in or out to accomplish the objective; ability to persist throughout shelf life; and, of course, the economics (Table 2.2).

Issues that must be considered in the potential use of antimicrobials in packaging include safety, efficacy, and adverse secondary effects.

Table 2.2.   Active packaging: antimicrobials proposed for or in commercial use.

- Antimicrobials
  - ▫ Contact
    - ▪ Silver salts
    - ▪ Acids
    - ▪ Antibiotics such as nisin
  - ▫ Noncontact
    - ▪ Allyl isothiocyanate
    - ▪ Chlorine dioxide
    - ▪ Ethanol
    - ▪ Natural spices such as oregano and carvacrol
    - ▪ Natural flavorants such as diacetyl

Ideally, the antimicrobials should be persistent in both long and short term and not encounter interferences due to interaction with the product or its processing. Adverse secondary effects reported have included off flavor, especially with natural spices, and color changes such as product darkening or bleaching.

Many packaging antimicrobials function only by direct contact with the microorganism, thus limiting the beneficial effect to food surfaces. Noncontact antimicrobials that are volatile function at a distance and so can be effective within the mass of the food product.

Regulatory concerns remain a major issue with all antimicrobials.

Antimicrobials may be married with or to the package material or structure by incorporation into or onto the plastic surface or other package material. The antimicrobial may be on the surface, in the structure, or adjacent to the product and should not interact with the package material.

Intelligent Packaging

Intelligent packaging senses change in the internal package or external environment and signals that change visually, audibly, or electronically. Examples of currently commercial intelligent packaging include maximum temperature experience; time/temperature integrators, which may be correlated to shelf life; and location sensors, the major current objective of radio frequency identification (RFID).

Intelligent packaging may also assist in computing the age of the contained product or the individual product identity for purposes of cook/reheat control in microwave or other domestic or food service heating devices. Intelligent packaging can identify and quantify concentrations of gases such as oxygen, CO_2, and ethylene. In theory, at least, the microbiological situation in terms of the total load or even the presence of a pathogen might be possible odorous end products of microbiological spoilage such as amines and/or sulfides which can be measured. Eventually, the quality of the food product might be signaled by direct measure or by inference.

Presently, information for retail distribution such as inventory control, pricing display and its changes, and automatic check out are nearly commercial intelligent packaging functions.

Issues associated with intelligent packaging include accuracy of information, repeatability, interpretation of the information by the receiver, application of information by the receiver, and the cost of the intelligence generated. In the future, it might be possible to convert the information into active packaging, that is, control the package function.

Nonthermal Preservation Technologies

Nonthermal technologies are food preservation with no or minimum input of heat. With nonthermal technologies, microbiological sterility is usually difficult to achieve, and hence the technologies may be applied instead for count reduction. Sometimes, nonthermal technologies can help control microbiological pathogens. Enzyme destruction is usually not achievable. Biochemical activity is generally not controlled by any of the mainstream nonthermal technologies.

Nonthermal preservation can be combined with minimal thermal processing, that is, mild heat, to achieve good levels of preservation usually for refrigerated distribution. Among the benefits of nonthermal preservation technologies are markedly reduced adverse thermal processing quality effects and extended shelf life (ESL), usually when linked with reduced temperature distribution for prolonged quality retention (Table 2.3).

The oldest and most highly publicized of the nonthermal preservation technologies is ionizing radiation, sometimes called "cold pasteurization." The objectives of ionizing radiation of food include pasteurization, pathogen reduction, insect reduction, and, in some situations, package material surface sterilization or, with very high doses, food sterilization.

Table 2.3. Nonthermal preservation technologies.

* Ionizing radiation
  * Gamma
  * X-ray
  * Electron beam
* Ultrahigh pressure or high-pressure processing
* Pulsed electrical
* Pulsed high-intensity light
* Ultrasonic
* Oscillating magnetic fields

Table 2.4. Ionizing radiation types.

* Penetrating radiation
  * Gamma ray from isotopic decay such as cobalt 60 or cesium 137
  * X-ray (controlled by operator)
* Low penetration
  * Electron beam (controlled by operator)
  * Ultraviolet—surface effect only

The well-publicized challenges of ionizing radiation include costs of the operations, consumer perception (often negative), package material interactions (leading to regulatory bans), adverse secondary effects such as flavor and/or nutritional value on the food, and regulatory restrictions. Interestingly, the most common consumer fear, radiation from the food, is virtually impossible to induce (Table 2.4).

Ultra-high pressure (UHP) has the objective of microbiological reduction, improved food functionality, and texturization. Additionally, UHP has been demonstrated to often assist in food processing such as opening mollusk or oyster shells. Issues associated with UHP processing include capital equipment and variable costs. Continuous equipment has yet to be engineered, and hence the process is not fast or with high output.

Pulsed electrical field (PEF) has the objective of chilled shelf extension for homogenous fluid foods. PEF offers little liquid quality change, but it is not suitable for particulate foods.

Pulsed high-intensity light is capable of surface decontamination and disinfection of transparent products, that is, water.

Ultrasound can provide some shelf life extension, but it is not for particulate foods.

Nonthermal preservation technology is usually a technically feasible food preservation method. The economics, even in theory, are questionable; that is, the initial capital investment, the variable costs, and/or the cost or benefit.

We do not know all there is to know about secondary effects, such as new chemical entities generated and stresses induced in package materials and structures which have been definitely identified for ionizing radiation and are inferred for UHP.

Today, there is relatively little commercial application for nonthermal preservation technologies. With ionizing radiation, which has been in and out of the commercial market for decades, there is consumer resistance and real or perceived safety concerns as well as economic issues.

UHP technology, which is a successful nonthermal technology, delivers very high-quality products such as guacamole, salsa, smoothies, fruit beverages, and ready meals for chilled distribution, but unfortunately it is still in relatively limited commercial application.

Combinations of preservation technologies can permit broader commercial applications as with minimum processing, hurdle technologies, enhancement of modified/controlled/reduced oxygen atmosphere, and linkages with packaging, such as aseptic and ESL.

Interface of Packaging and Nonthermal Processing

Quality and safety retention depend on the initial quality of the food product, distribution conditions, and packaging to obviate microbiological recontamination and to control water vapor transmission and oxygen.

Nonthermal processing in sequence with packaging requires transfer of the food into the package by aseptic or other means under reduced or elevated oxygen. After packaging, nonthermal processing generally requires hermetic closure before processing and maintenance of hermetic closure throughout the process and distribution. Nonthermal process also may be conducted while in package prior to hermetic closure.

Questions that might be asked at this time include the following: Why are combinations of active packaging and nonthermal processing interesting? What is driving these interests? and, How are active packaging structures influenced by nonthermal technologies?

We shall attempt to describe the various visible interfaces between active packaging and nonthermal processing.

Active Packaging: Moisture Control

Ionizing radiation probably exerts adverse effects on desiccants and absorbent pad active components. Further, radiation affects liquid purge or condensate from relative humidity. Free radicals form and produce secondary microbicidal effects as well as increase or decrease adverse biochemical effects. Increased microbicidal effects on the purge may be concentrated in one site. With UHP, the effect of increase or decrease in water content, if any, is not quantified. Might UHP affect a desiccant or absorbent pad active component? Little or no perceptible interaction has been noted with PEF because it is a prepackaging process or with high-intensity pulsed light.

Oxygen Scavengers/Absorbers

Ionizing radiation can have an effect on scavengers themselves, whether they are ferrous iron, unsaturated hydrocarbons, or polyamides. The effect should be different with different oxygen concentrations. Secondary effects may arise, and there may be increases in product biochemical oxidation rate with increased oxygen levels.

No reports have been noted on the effects of UHP processing on oxygen scavenger systems. On the other hand, might there be effects of oxygen scavengers on UHP efficacy? Oxygen removal should assist in extending biochemical shelf life during distribution.

Ionizing Radiation

Ionizing radiation levels probably increase adverse biochemical reactions and ozone associated with active oxygen generation, which adversely reacts with food product. On the other hand, possible increased microbicidal effects might result.

UHP probably has no direct interactions with elevated oxygen packaging. Incidentally, decreases of biochemical shelf life occur as a result of elevated oxygen.

CO_2 Control: Ionizing Radiation and UHP

Does elevated CO_2 assist in suppressing oxidation effect, probably only by displacing oxygen? Does elevated CO_2 complement antimicrobial effect, probably only by reducing pH? What, if any, are the effects of reduced CO_2 levels? Does CO_2 complement or compromise the microbicidal effects of UHP?

Antimicrobials

Do ionizing radiations suppress, enhance, or otherwise interfere with chemical antimicrobial effects? Might ionizing radiation retard the movement passage or permeation of antimicrobials? Does UHP processing complement microbicidal effects by possibly facilitating the movement of antimicrobial or alter the food product structure sufficiently to facilitate movement? Or might UHP processing interfere with antimicrobial effects by retarding migration of the antimicrobial?

Modified Atmosphere/Reduced Oxygen Packaging

Modified atmosphere/reduced oxygen (MAP/RO) packaging is not a new nonthermal process. Both, independently and combined, are widely applied commercially to extend refrigerated shelf life of fresh and minimally processed foods: fresh produce, fresh meat, fish, produce, and so on. MAP/RO often employs controlled gas (oxygen) permeability package structures to function optimally or nearly optimally.

MAP/RO can be enhanced by active packaging such as moisture control in the package structure. MAP functions best in aqueous products. In excess moisture conditions, the probability of surface microbiological growth is higher, and hence MAP can display more effectiveness. Oxygen control usually requires passage in of controlled volumes of atmospheric air and assurance of specific levels of oxygen within the package.

CO_2 control often requires CO_2 generation to aid MAP functionality. Active packaging can in part control the ratio of CO_2 to oxygen to optimize the MAP effect. Removal of excess CO_2 in some situations can retard adverse secondary effects.

MAP/RO might enhance the microbicidal/microbistatic effects of antimicrobials. The antimicrobial effects of MAP permeating throughout products are complementary to contact antimicrobials. Alteration of pH-controlling antimicrobials is enhanced by MAP action.

Hurdle Technologies

If little or minimum heat is applied, hurdle technologies might be considered as classical nonthermal processes. Hurdle is defined as combinations of multiple technologies in low doses to function in synergy with each other to maximize preservation. Synergies may be derived from active packaging.

Intelligent Packaging

Might maximum temperature indicators be interfered with by ionizing radiation or UHP processing? With time/temperature integrators, there is a relatively high probability of interference by ionizing radiation or high-pressure processing even to the point of disrupting the active agent.

Radio Frequency Identification Devices
RFID is being proposed and even applied for signaling many variables such as location, microbiological content/activity, and/or quality values. Because RFID depends on sensitive electrical signals, a relatively high probability exists of interference or even disruption by ionizing radiation or high-pressure processing (or, for that matter, the presence of moisture).

Conclusions

Some active packaging has a high probability of wider applications for moisture and oxygen control and CO_2 generation and has some promise in antimicrobials and self-heating packages.

Much intelligent packaging holds promise for commercial application in time/temperature integration as a surrogate for shelf life indication, in location indicators for inventory control, in price/check-out signalers, in microwave oven control, and eventually to be converted into active packaging.

UHP processing may be expected to be more widely applied to high-value food products to deliver higher quality refrigerated foods. There should be much increased application of hurdle technologies as they are better quantified. Obviously, MAP/RO will continue to grow from their present broad base.

MAP packaging coupled with oxygen/CO_2 control offers major synergistic benefits. Major marriages of active packaging and nonthermal preservation are possible. Logical relationships exist between hurdle technologies and active packaging. Other connections of active and intelligent packaging and nonthermal processing are not highly visible. Some linkages appear to suggest mutual interferences. Care must be exercised in attempting forced marriages of nonthermal preservation and active packaging.

Where Do We Go From Here?

Needed for active packaging are more comprehensive reduced oxygen control, more functional antimicrobials with fewer secondary effects, and reliable persistent CO_2 sourcing.

Needed for intelligent packaging are sensors to actuate RFID and RFID devices with lower environmental sensitivity.

In nonthermal processing, ionizing radiation must exhibit fewer adverse secondary side effects perhaps by the incorporation of free radical acceptors. UHP needs continuous shorter-time processing and/or out-of-package processing followed by aseptic packaging. Hurdle technologies need quantification of the effects to facilitate application of each contribution.

Needs of (desires for) the packaging–nonthermal processing interfaces include active packaging agents that offer solutions to nonthermal processing issues such as active free radical acceptors in package structures for ionizing radiation; in-package-structure oxygen removal to reduce oxygen to near zero; high-pressure-sensitive antimicrobial release/oxygen scavenger systems; and greater concentration of active packaging agents from channel structures.

The potential is so great for some nonthermal processes and for some active and intelligent packaging that the three disciplines demand a total systems approach to link them and gain the synergies that appear to be near.

References

Brody, Aaron L., E. R. Strupinsky, and Lauri R. Kline, 2002, *Active Packaging for Food Applications*, CRC Press, Boca Raton, FL.

Cha, Dong, and Manjeet Chinnan, 2004, Biopolymer-based antimicrobial packaging: a review. *Critical Reviews in Food Science and Nutrition*, 44: 223–237.

Han, Jung, 2005, *Innovations in Food Packaging*, Elsevier, Amsterdam.

Miltz, Joseph, 2004, Presentation at BARD International Workshop on Active and Intelligent Packaging, Sheperdston, WV.

Versteylen, Sayandro, 2005, Presentation at Food Processors Association Meeting on Active Packaging, Chicago, IL.

Chapter 3

EDIBLE COATINGS CONTAINING
BIOACTIVE ANTIMICROBIAL AGENTS

Seacheol Min and John M. Krochta

Introduction

Edible biopolymer films are manufactured as either food coatings or stand-alone film wraps, layers, casings, and pouches (Miller and Krochta 1997). Edible coatings are formed on foods by dipping, spraying, or panning.

The purposes of using the films and coatings, depending on the food, are to inhibit the migration of oxygen, carbon dioxide, aromas, oil, and moisture; enhance the appearance of the food; improve mechanical integrity; and carry nutritional supplements and food additives (Kester and Fennema 1986; Krochta and Mulder-Johnston 1997). An edible film coating, acting as an efficient moisture, oxygen, or aroma barrier, can reduce packaging requirements and waste. For example, the barrier characteristics of an edible film coating may allow for conversion from a multilayer plastic package to a single-component recyclable package (Miller and Krochta 1997). Sensory appeal is an important functionality of an edible film/coating. The sensory characteristics of edible films and coatings such as color, gloss, transparency, roughness, and sticking can be modified depending on the purposes (Debeaufort, Quezada-Gallo, and Voilley 1998). Edible films and coatings may also carry food additives that improve general coating performance (e.g., strength, flexibility, adherence); enhance product color, flavor, and texture; and control microbial growth during marketing and even after a packaging is opened (Cuppett 1994). Thermoforming, over-wrapping, and shrink-wrapping of foods with edible films are also conceivable (Krochta and Mulder-Johnston 1997).

Edible films and coatings based on polysaccharides, proteins, and/or lipids can be made antimicrobial by incorporating approved antimicrobial agents. For example, effective amounts of potassium sorbate, nisin, or lysozyme can be incorporated into whey protein films (Han 2000). Examples of those agents are organic acids (e.g., acetic acid, propionic acid, sorbic acid), bacteriocins (e.g., nisin, pediocin, lacticin), and enzymes (e.g., lysozyme, lactoperoxidase systems [LPOSs]).

This chapter describes (1) the advantages of using edible films and coatings containing antimicrobials, (2) biopolymers to form edible film and coating matrices, (3) bioactive antimicrobials used in edible films and coatings, (4) requirements for producing the films and coatings, (5) methods available for testing antimicrobial activity of the films and coatings, and (6) diffusion of the antimicrobials in the films and coatings. A combination use of the technologies of bioactive antimicrobial coating and nonthermal food processing is also proposed in this chapter.

Advantages

Edible films are effective carriers of antimicrobials (Cuppett 1994; Min, Harris, and Krochta 2005a). We can simply treat foods with these compounds by direct applications such as dipping, dusting, or spraying on the surface of the food. However, when antimicrobials are added by direct food-surface application, they quickly diffuse away from the surface and are rapidly diluted and/or react with food components. When the antimicrobials are incorporated in a film/coating matrix, their migration to the surface of the film and food can be controlled in such a way that they are available at a desired concentration for an extended period of time. Thus, smaller amounts of antimicrobial agents would be needed to achieve a targeted shelf life compared with dipping, dusting, or spraying antimicrobials on the surface of the food. The diffusion of the incorporated antimicrobial can be controlled by controlling film composition (e.g., ratio of film base material to plasticizer) and/or pH. The use of multilayered edible films was proposed to achieve appropriate controlled release of the antimicrobial to the food surface (Han 2000).

Base Materials of Edible Films and Coatings

Polysaccharides, proteins, and lipids are the major classes of compounds that form a continuous film or coating matrix. The choice of the base material depends on the specific application of the films and coatings (i.e., the type of food product and main deterioration mechanisms) (Cuq, Gontard, and Guilbert 1995). For example, the type of coating material applied alters fruit and vegetable skin permeability to oxygen and carbon dioxide. The control of gas exchange results in improved control of the ripening of fruits (Debeaufort, Quezada-Gallo, and Voilley 1998). Water-barrier effectiveness of coatings is also desirable to retard food degradations.

If edible coatings have a particular taste or flavor, their sensory characteristics must be compatible with those of the food application to prevent detection of the use of coating during consumption (Biquet and Labuza 1988).

The materials that have received attention for edible film and coating use are summarized, along with their functions, in Table 3.1.

Antimicrobial Agents

The antimicrobials used in edible films and coatings and the resulting antimicrobial activities of the films and coatings are listed in Table 3.2. The choice of the antimicrobial is influenced by the compatibility of the agent with the film base material and the method used for producing films (Han and Floros 1997). Thermal processing methods of producing films or packages, such as extrusion and injection molding, would only be used with thermally stable antimicrobials (Appendini and Hotchkiss 2002).

Lysozyme

Lysozyme is of interest for use in foods because it is a naturally occurring enzyme that is produced by humans and many animals (Gill and Richard 2000). Lysozyme inhibits bacterial growth by the hydrolysis of glucosidic linkages in the peptideglycan of cell walls. Specifically, lysozyme hydrolyzes the β-1,4 linkages between *N*-acetylmuramic acid and 2-cetyl-amino-2-deoxy-D-glucose residues in bacterial cell walls, resulting in cell lysis (Shah 2000). Lysozyme is usually active against several gram-positive bacteria such as *Listeria monocytogenes* (Losso, Nakai, and

Table 3.1. Materials for edible biopolymer films and coatings and their principal functions.

	Material	Principal Function	References
Polysaccharide	Starch, hydroxypropylated starch, methyl cellulose (MC), carboxy methyl cellulose (CMC), hydroxyl propyl cellulose (HPC), hydroxyl propyl methyl cellulose (HPMC), amylose, pectin, guar gum, locust bean gum, tara gum, pullulan, alginate, carrageenan, and chitin-chitosan	Oxygen- and carbon dioxide-barrier properties (effective control of gas exchange and creation of a modified atmosphere inside), but poor moisture-barrier properties (water loss from coated foods) High relative humidity (RH) conditions can cause swelling of the matrix, resulting in increased permeability Oil-barrier property	Kester and Fennema (1986); Aoyama *et al.* (1993); Conca and Yang (1993); Krochta and Mulder-Johnston (1997); and Krochta, Min, and Han (2005)
Protein	Whey protein, corn zein, wheat gluten, soy protein, casein, collagen, gelatin, maize, β-lactoglobulin, egg white protein, keratin, fish myofibrillar protein, peanut protein, cottonseed protein, and rice bran protein	Carriers of food ingredients including antimicrobial and antioxidant An increase in cross-linking, crystallinity, density, orientation, or molecular weight decreases polymer permeability and thus increases moisture-barrier property. Heat	Miller and Krochta (1997); Perez-Gago and Krochta (2001); Krochta, Min, and Han (2005); Min, Harris, and Krochta (2005a)

		denaturation of whey proteins at specified conditions results in films with improved barrier and mechanical properties that are insoluble in water Labeling required for consumers who have wheat gluten intolerance, milk protein allergies, or lactose intolerance	
Lipid	Beeswax, carnauba wax, candelilla wax, paraffin wax, sisal paraffin, carnauba wax, rice bran wax, sugarcane, acetylated monoglycerides, and shellac	Water loss reduction, improved appearance by imparting a subtle shine, and creation of a modified atmosphere	Kester and Fennema (1986); Baldwin et al. (1999); Krochta, Min, and Han (2005)
Composites	Hydrolyzed lecithin, acrylate and methacrylate polymers, polymers of vinyl acetate, organic latex, and copolymers (vinyl acetate, acrylate, ethyl acrylate, and propyl acrylate)	Low water vapor transmission and adequate permeability to oxygen and carbon dioxide	Baldwin (1994); McHugh and Krochta (1994); Wong, Camirand, and Pavlath (1994); Park (1999); Coma et al. (2001)
	Mixtures of sucrose esters of fatty acids, CMC sodium salt, and mono/diglycerides of fatty acids: TAL Pro-long (Courtaulds Group, London, UK) and Semperfresh™ (United Agriproducts, Greeley, CO, USA)	Composite films of fatty acids in polysaccharide films: decreased high moisture transfer rate and reduced increase in gas permeability at high RH in polysaccharide films and coatings	
	Composites: CMC/sucrose fatty acid esters/mono- and diglycerides, pullulan/sorbitol/sucrose fatty acid ester, alginic acid/casein/acetylated monoglyceride	Composite films of wax or other lipid materials in protein films: Lowered water vapor permeability of protein films and coatings	

Table 3.2. Bioactive antimicrobial agents incorporated in biopolymer films and coatings for their antimicrobial activity.

Bioactive Antimicrobial Agent	Biopolymer Film/ Coating Matrix	Microbial Inhibition	Applied Food	References
Pediocin	Cellulose	*L. monocytogenes*	Meat	Ming *et al.* (1997)
Lysozyme	Cellulose triacetate	*Micrococcus lysodeikticus*		Appendini and Hotchkiss (1997)
Lysozyme, nisin	Soy protein isolate, corn zein	*Lactobacillus plantarum*		Padgett, Han, and Dawson (1998)
Nisin	Methylcellulose (MC), hydroxy propyl methyl cellulose (HPMC)	*S. aureus, L. monocytogenes*		Cooksey (2000)
Nisin	HPMC	*L. innocua, S. aureus*		Coma *et al.* (2001)
Nisin	Whey protein isolate, (WPI) soy protein isolate, egg albumen, wheat gluten	*L. monocytogenes*		Ko *et al.* (2001)
Chitosan (1% (w/v) of film-forming solution)	Chitosan	*L. monocytogenes* (200–250 CFU/Petri dish), *L. innocua*		Coma *et al.* (2002)
Grapefruit seed extract (GSE)	Na-alginate and κ-carrageenan	*Micrococcus luteus*		Cha (2002)
Nisin (4% (w/w) of film)	Soy based	*L. monocytogenes* (brain heart infusion agar [BHI]: 6 log reduction; turkey bologna: 1 log reduction)	Turkey bologna	Dawson *et al.* (2002)

Nisin (5×10^4 IU/mL of film-forming solution)	HPMC	L. monocytogenes, S. aureus		Sebti, Ham-Pichavant, and Coma (2002)
Nisin (205 IU/g protein) +malic acid	Soy protein isolate	L. monocytogenes, E. coli O157:H7, S. gaminara (5.5, 6.8, 3.0 log CFU/mL, respectively)		Eswaranandam, Hettiarachchy, and Johnson (2004)
Bacteriocin-like substance (BLS) (0.32%)	Fermented soybean meal with Bacillus subtilis	E. coli		Kim et al. (2004)
Lysozyme (780 µg/cm² film) + disodium EDTA·2H_2O (520 µg/cm² film)	Crude exopolysaccharides (59% pollulan)	E. coli		Kandemir et al. (2005)
Lysozyme (25 mg/g coating solution)	WPI	L. monocytogenes (2.4 log CFU/g reduction)	Smoked salmon	Min et al. (2005)
Lactoperoxidase system (0.15 g/g film)	WPI	Salmonella enterica, E. coli O157:H7 (4 log CFU/cm² reduction)		Min, Harris, and Krochta (2005a)
Lactoperoxidase system (40 mg/g coating)	WPI	L. monocytogenes (3 log CFU/g reduction)	Smoked salmon	Min, Harris, and Krochta (2005c)
Garlic oil (100 µg/g chitosan)	Chitosan	S. aureus, L. monocytogenes, Bacillus cereus		Pranoto, Rakshit, and Salokhe (2005)
Garlic oil [0.4% (v/v)]	Alginate	S. aureus, B. cereus		Pranoto, Salokhe, and Rakshit (2005)
Oregano essential oils (1–2% of film)	Chitosan	L. monocytogenes, E. coli (3.6–4.0, 3.0 logs, respectively)		Zivanovic, Chi, and Draughon (2005)

35

Charter 2000). Min *et al*. (2005) clearly demonstrated that the application of whey protein films and coatings incorporating lysozyme inhibits the growth of *L. monocytogenes* present in smoked salmon.

Lactoperoxidase System

The LPOS is a natural antimicrobial system in human secretions such as saliva, tear fluid, and milk (Kussendrager and van Hooijdonk 2000). The use of LPOS has been suggested as a preservative in foods and pharmaceuticals (Bosch, Van Doorne, and De Vries 2000). The lacto-peroxidase (LPO) in LPOS catalyzes the oxidation of thiocyanate ion (SCN^-), generating oxidizing products such as hypothiocyanite ($OSCN^-$) and hypothiocyanous acid (HOSCN), which inhibit the growth of microorganisms by the oxidation of sulfydryl (SH) groups of microbial enzymes and other proteins (Kussendrager and van Hooijdonk 2000). The structural damage of microbial cytoplasmic membranes by the oxidation of SH groups results in the death of the cells (Reiter and Harnulv 1984).

LPOS-incorporated whey protein films may be used for inhibiting the growth of *L. monocytogenes*, *Escherichia coli* O157:H7, and *Salmonella* already present on smoked salmon and roasted turkey as well as for controlling their growth, resulting in contamination, after wrapping or coating of the food products (Min, Harris, and Krochta 2005b,c).

Nisin

Nisin is one of the most investigated bioactive antimicrobials used in antimicrobial film and coating studies. Nisin inhibits the growth of microorganisms by interacting with sulfur-containing compounds in the microbial membrane, disrupting their semipermeable function, and causing lysis of the microbial cells (Natrajan and Sheldon 2000). Nisin has been accepted for food use by regulatory authorities in some countries such as Japan (Brody, Strupinsky, and Kline 2001). Nisin films and coatings inhibit the growth of *Staphylococcus aureus*, *L. monocytogenes*, *Listeria innocua*, *E. coli* O157:H7, and *Salmonella gaminara* (Table 3.2).

Pediocin

Pediocin is another bacteriocin that has been found to be effective against *L. monocytogenes*. A pediocin-coated cellulose casing was developed as a means of inhibiting the growth of *L. monocytogenes* in meat (Ming *et al.* 1997).

Chitosan

Chitosan, listed as an edible coating material in Table 3.1, has an inherent antimicrobial activity. Charged amines of chitosan interact with negative charges on the cell membrane of microorganisms. This may cause leakage of intracellular constituents and result in the eventual death of microorganisms (Goldberg, Doyle, and Rosenberg 1990). Chitosan films and coatings have been shown to inhibit the growth of *L. monocytogenes* on an agar medium (Coma *et al.* 2002).

Essential Oils

Essential oils and their components are becoming popular as antimicrobial agents. The essential oils of *Melaleuca alternifolia* (Carson and Riley 1995), *Pelargonium* (Lis-Balchin *et al.* 1998), Sardinian *Thymus* (Cosentino *et al.* 1999), such as carvacrol and thymol, *Senecio graveolens* (Compositae) (Perez, Agnese, and Cabrera 1999), *Picea excelsa* (Canillac and Mourey 2001), and *Ocimum gratissimum* leaf (Orafidiya *et al.* 2001) have been reported as potential antimicrobial agents (Cha and Chinnan 2004).

Allyl Isothiocyanate

A major flavoring component of wasabi and mustard, identified as allyl isothiocyanate (AIT), has been reported as an effective antimicrobial agent (Cha and Chinnan 2004). AIT reduced the number of *E. coli* O157:H7 in fresh ground beef during refrigerated and frozen storages (Nadarajah, Han, and Holley 2005).

Other Plant Extracts

Antimicrobial activities of plant-origin compounds, including extracts of grapefruit seed, cinnamon, allspice, clove, thyme, rosemary, onion, garlic, radish, mustard, horseradish, and oregano, have been demonstrated (Brody, Strupinsky, and Kline 2001; Suppakul *et al.* 2003). Among those, grapefruit seed extract (GSE) has been used as a food additive in several countries (Cha and Chinnan 2004). It contains naringin, ascorbic acid, hesperidins, and various organic acids, such as citric acid, and has a wide spectrum of microbial growth-inhibition properties. The antimicrobial activity of GSE is known to be stable at high temperature, which makes it suitable for a commercial production of films by extrusion (Cha and Chinnan 2004).

Development Factors

Factors to be considered in the development of an edible film or coating incorporating an antimicrobial agent include the following:

1. Contact of the antimicrobial agent with food should be approved. For example, lysozyme and nisin are substances that are generally recognized as safe (GRAS), in accordance with Food and Drug Administration (FDA) Title 21 of the Code of Federal Regulations (21 CFR).
2. The edible antimicrobial film or coating needs to be cost-effective. If cost is too high, commercialization may be viable for only high-value food products (e.g., sea foods).
3. The polarity, ionic strength, molecular weight, and solubility of the antimicrobials should be considered, because they affect the rate of diffusion in films and coatings (Suppakul *et al.* 2003). When an antimicrobial polypeptide is incorporated into a protein film or coating, interaction with the protein of the film or coating must be considered. Nisin was found to interact differently with proteins of different films (Cagri, Ustunol, and Ryser 2004).
4. Characteristics of the food need to be considered in the manufacture. For example, the pH of a food product influences the growth rate of target microorganisms and changes the activity of antimicrobials (Han 2000). Diffusion of antimicrobials may be affected by the water activity of foods.
5. The activity of antimicrobial films and coatings may be affected by the storage temperature, because diffusion is known to be directly related to temperature. Higher diffusion rate occurs at higher temperature.
6. Activity of the incorporated antimicrobial compound should be maintained during the manufacture and shelf life of the film/coating-protected food product.

7. Adhesion, involving interactions between coating molecules and food-surface molecules, should be carefully considered for a successful application of edible films and coatings to foods (Miller and Krochta 1997). The characteristics of the base material, the concentration of the material in the film/coating-forming solution, holding temperature and time for solution heating (if applicable), and evaporation rate of the solvent (if used) are important processing parameters influencing adhesion (Miller and Krochta 1997).
8. Aroma- and oxygen-barrier properties, tensile properties, and the color of edible films and coatings may be affected by the incorporation of antimicrobials. Depending on the desired applications, the changes may be either positive or negative.

Activity Measurement

Methods for evaluating edible antimicrobial film and coating activity may be divided into three types (Min and Krochta 2005):

Type 1: antimicrobial activity of film- or coating-forming solutions on microbial agar media;
Type 2: antimicrobial activity of films on microbial agar media;
Type 3: measurement of the activity of films or coatings on foods.

Recently used methods to evaluate the effect of edible antimicrobial films and coatings on the inhibition of microorganisms are shown in Table 3.3.

Inhibition Zone Test (Type 1)

A lawn of a target microorganism on agar plates (8.5 cm in diameter) is formed by overlaying a 0.5–0.8% (w/v) agar seeded with a target microorganism. The microbial density of the lawn generally ranges from 10^4 to 10^6 colony-forming units (CFU)/plate. A film-forming solution is dropped on the lawn of the target microorganism. Film-forming solutions at different concentrations may be tested, depending on the experimental design. Dishes are refrigerated at 4°C for 3 h, to allow diffusion of the antimicrobial agent, and then incubated at desired conditions. After incubation for a desired time, the antimicrobial activity

Table 3.3. Methods for testing effects of bioactive antimicrobial films and coatings against microorganisms.

Methods for Testing Inhibition	Base Materials for Films/Coatings	Target Microorganisms	Antimicrobial Agents	Media/Food Used for Tests	Effects	References
Inhibition zone assay, disc-covering test	Hydroxy-propyl methyl cellulose (HPMC)	*M. luteus, L. mono-cytogenes, S. aureus*	Nisin	Nutritive agar, tryptose agar	All tested bacteria were inhibited	Sebti, Ham-Pichavant, and Coma (2002)
Inhibition zone assay	Alginate	*S. aureus, B. cereus*	Garlic oil	Mueller Hinton agar	All tested bacteria were inhibited	Pranoto, Salokhe, and Rakshit (2005)
Disc-covering test	Soy protein isolate, corn zein	*L. plantarum, E. coli*	Lysozyme, nisin, ethylene-diaminet etraacetic acid (EDTA)	*Lactobacilli* MRS agar, trypticase soy agar	Inhibition of bacterial growth	Padgett, Han, and Dawson (1998)
Disc-covering test	HPMC	*M. luteus, L. innocua, S. aureus*	Nisin	Nutritive agar, tryptose agar	All tested bacteria were inhibited	Coma *et al.* (2001)
Disc-covering test, plate-counting method	Chitosan	*L. innocua* 430, *L. monocy-togenes*	Chitosan	Tryptose broth and agar, Emmental cheese	Bactericidal activity with 1% (w/v) chitosan in the film-forming solution	Coma *et al.* (2002)

40

Method	Substrate	Antimicrobial	Organism	Application	Results	Reference
Direct inoculation method	Soy protein isolate spray-dried wheat gluten, egg albumen protein, whey protein isolate	Nisin	*L. monocytogenes*	Films	Inhibition increased as the nisin concentration increased. ~2 log reduction (inoculum size: ~5 log CFU/mL) by 160IU nisin in whey protein isolate films	Ko et al. (2001)
Direct inoculation method, plate-counting method	Soy protein isolate	Lauric acid, nisin	*L. monocytogenes*	Turkey bologna	Films containing both lauric acid and nisin completely inhibited cells from a 10^6 CFU after 8 h of exposure to 1% peptone medium at 22°C. Films with lauric acid, nisin, or both reduced the cell number of turkey bologna by 1 log after 21 days at 4°C	Dawson et al. (2002)

(continued)

41

Table 3.3. Methods for testing effects of bioactive antimicrobial films and coatings against microorganisms. (*Continued*)

Methods for Testing Inhibition	Base Materials for Films/Coatings	Target Microorganisms	Antimicrobial Agents	Media/Food Used for Tests	Effects	References
Plate-counting method	Gelatin	*Brochothrix thermosphacta, Lactobacillus sakei, Leuconostoc mesenteroides, L. monocytogenes, Salmonella typhimurium*	Lysozyme, nisin, EDTA	Ham, bologna	Coating treatment resulted in bactericidal effects up to 4 log CFU/cm^2 on *B. thermosphacta, Lactobacillus sakei, Leuconostoc mesenteroides, L. monocytogenes,* and *S. typhimurium* and inhibited the growth of these microorganisms in bologna during the 4 weeks of storage at 8°C	Gill and Richard (2000)

is determined by measuring the size of zone of inhibition. The larger the zone of inhibition, the stronger the antimicrobial activity.

Disc-Covering Test (Type 2)

A lawn of a target microorganism is formed as described in *Inhibition Zone Test (Type 1)*. Circular film discs (0.5–2.0 cm in diameter) containing antimicrobial agents are placed on the preformed lawn. After appropriate incubation, the clear zone of inhibition around the film disc in the bacterial lawn is visually examined, and the size of the clear zone is measured to the nearest 1 mm (Coma *et al.* 2002; Eswaranandam, Hettiarachchy, and Johnson 2004; Min, Harris, and Krochta 2005a,c). The antimicrobial activity correlates with the size of inhibition zones. This test simulates what would happen if the antimicrobial film were wrapped or the coating were formed over a previously contaminated food (Min, Harris, and Krochta 2005a). An example of results from the disc-covering test is shown in Figure 3.1(A).

Disc-Surface-Spreading Test (Type 2)

A film disc (0.5–2.0 cm in diameter) is placed on top of a solidified agar medium. A microbial inoculum is then spread over the plate (10^4–10^6 CFU/plate) and the plate incubated (Halek and Garg 1989; Min, Harris, and Krochta 2005a,c). After incubation, the antimicrobial activity is determined as described in *Disc-Covering Test (Type 2)*. This test simulates post-contamination by microorganisms on the surface of edible films/coatings previously placed or formed on a food surface (Min, Harris, and Krochta 2005a). An example of the results from the disc-surface-spreading test is demonstrated in Figure 3.1(B).

Direct Inoculation Test (Type 2)

A certain volume (e.g., 15 μL) of a microbial suspension (10^5–10^9 CFU/mL) is inoculated on a film disc. After incubation of the inoculated film disc for a desired time, a visual observation or an epifluorescence test (Coma *et al.* 2002) is conducted on the disc to examine its antimicrobial activity.

(A)

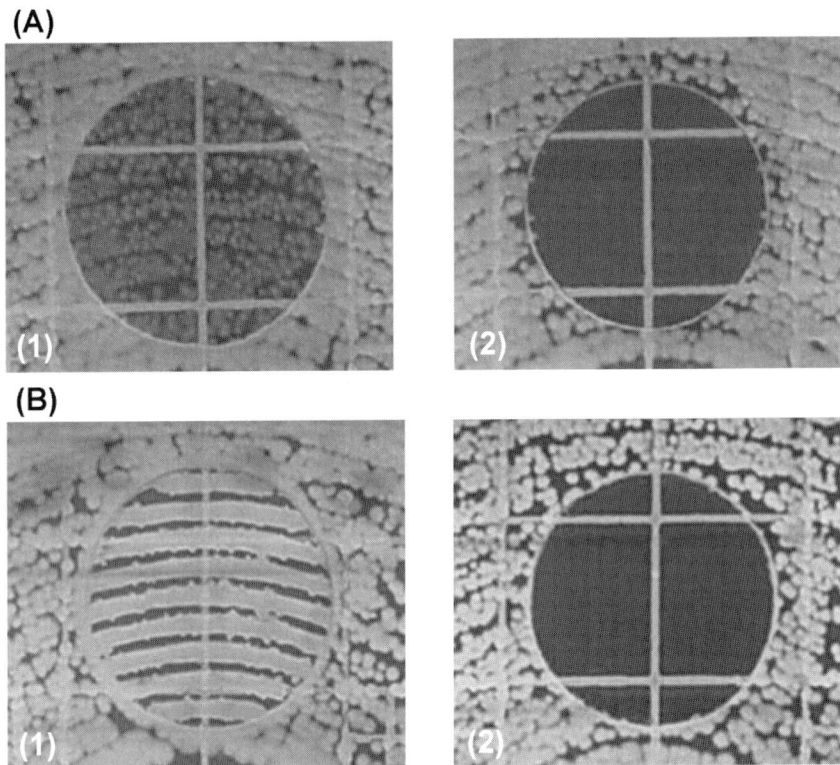

(B)

Figure 3.1. Effects of the whey protein isolate (WPI) films incorporating the lactoperoxidase system (LPOS) at 0.15 g/g film (dry basis) on the growth of *Escherichia coli* O157:H7 (10^3 colony-forming units [CFU]/cm^2) on tryptic soy agar. Panels (A) and (B) show results obtained by the disc-covering test and the disc-surface-spreading test, respectively. Images (1) and (2) in (A) and (B) are of control WPI films without the LPOS and the films incorporating the LPOS, respectively.

Plate-Counting Method (Type 3)

Food samples are prepared with a uniform dimension. A coating is formed on the surface of each sample by either applying a coating solution or wrapping it with a film after or before inoculating with a target microorganism (10^3–10^5 CFU/g). The inoculation can be done by spotting and spreading. Inoculated samples are incubated for a desired time at desired conditions and sampled every selected period of time. Samples are blended with a buffer solution or 0.1% peptone water by a stomacher. From the resulting homogenate, serial dilutions are prepared and the appropriate dilutions spread-plated onto agar plates. The plates are incubated as desired. The survival of microbial cells is measured by their colony-forming ability on the plates. Reduction of growth rate and reduced maximum growth population indicate improved microbial safety, and an extended lag period indicates a prolonged shelf life with microbial quality assurance (Chen, Weng, and Chen 1999; Yaman and Bayindirh 2001; Min *et al.* 2005; Min, Harris, and Krochta 2005b).

Headspace Gas Composition Assay (Type 3)

The rate of increase in CO_2 in the headspace of sealed glass jars can be used as a measure of mold growth. Food samples, which are coated or wrapped before or after inoculation with a target number of fungal spores (e.g., 10^4 spores), are incubated in conditioned glass jars. The gas composition of headspace in each glass jar is measured during storage using gas chromatography (Weng and Hotchkiss 1991, 1992).

Antimicrobial Diffusion

Diffusion coefficient (*D*) values for antimicrobials in films must be determined to understand diffusion phenomena in polymer matrices and to assess the ability of selected polymers to act as antimicrobial carriers (Franssen, Rumsey, and Krochta 2004).

Various chemical and physical factors may be involved in the diffusion of an antimicrobial in a biopolymer matrix. Chemical factors include hydrogen bonds, ionic bonds, ionic osmosis, electrostatic interactions, and hydrophobic interactions between the antimicrobial and the film/coating matrix (Suppakul *et al.* 2003). Physical factors include the conformation (conformational change), porosity, and tortuosity of the matrix; the partition coefficient between the film/coating and food (*K*); and the steric hindrance and frictional resistance of the incorporated antimicrobial (Kuu, Wood, and Roseman 1992).

Examples of mathematical models used to quantify the diffusion in films are shown in Table 3.4. The solutions, derived from Fick's second law of diffusion, have adequately predicted *D* values of antimicrobials in edible films (Ouattara *et al.* 2000; Franssen, Rumsey, and Krochta 2004).

Both experimental diffusion coefficient (D_e) and the diffusion coefficient in water at infinite dilution (D_0, diffusion for a moving substance) have been used to characterize the releases of α-chymotrypsin, β-lactoglobulin, bovine serum albumin (BSA), and γ-globulin from lactitol-based cross-linked hydrogel (Han *et al.* 2000). The diffusion coefficient for lysozyme diffusion in agarose gel was determined using Fick's second law and a

Table 3.4. Mathematical equations used to quantify diffusion of antimicrobials in biopolymer films.

Diffusant	Biopolymer Film Matrix	Equations Applied[1]	References
α-Chymotrypsin, β-lactoglobulin, bovine serum albumin (BSA), and γ-globulin	Lactitol-based cross-linked hydrogel	$D_0 = \dfrac{kT}{6\pi\mu R_h}$, $\ln\left(1 - \dfrac{M_t}{M_\infty}\right) = \ln\dfrac{8}{\pi^2} - \dfrac{D_e \pi^2 t}{4L^2}$ $(M_t/M_\infty > 0.4)$	Han et al. (2000)
Lysozyme	Agarose gel	$\dfrac{\partial c(x,t)}{\partial t} = D\,\dfrac{\partial^2 c(x,t)}{\partial x^2}$ (Fick's second law) $\dfrac{\partial c(x,t)}{\partial t} = \dfrac{D}{RT}\,c\,\dfrac{\partial}{\partial x}\left(\dfrac{\partial \mu(x,t)}{\partial x}\right)$	Mattisson et al. (2000)
Acetic acid, propionic acid	Chitosan	$\dfrac{M_t}{M_\infty} = 1 - \dfrac{8}{\pi^2}\sum_{n=0}^{\infty}\dfrac{1}{(2n+1)^2}\exp\left\{-D(2n+1)^2\pi^2 t/h^2\right\}$ $D = \dfrac{0.049h^2}{t_{0.5}}$ $(M_t/M_\infty = 0.5)$	Ouattara et al. (2000)
Potassium sorbate	Whey protein isolate	$\dfrac{M_t}{M_\infty} = 1 - \dfrac{8}{\pi^2}\sum_{n=0}^{\infty}\dfrac{1}{(2n+1)^2}\exp\left\{-(2n+1)^2\pi^2 T\right\}$, where $T = \dfrac{D_p t}{L^2} + \dfrac{8D_p}{\pi^4 D_s}\sum_{n=0}^{\infty}\dfrac{1}{(2n+1)^4}\exp\left[-(2n+1)^2\pi^2 D_s t/L^2\right] - \dfrac{D_p}{12D_s}$ $(n=10)$	Ozdemir and Floros (2003)

[1]Mathematical solutions are not shown here.

D_0, diffusion coefficient in water at infinite dilution; D_e, experimental diffusion coefficient; k, Boltzmann constant; T, absolute temperature; μ, solvent viscosity; M_t, amount released at time t; M_∞, amount released at time infinity; L, gel half thickness; c, concentration; μ, chemical potential; h, film thickness; $t_{0.5}$, the time at M, $0.5M_\infty$; D_s, solvent diffusion coefficient; D_p, diffusion coefficient at equilibrium solvent absorption.

general model derived from Fick's law accounting for chemical potentials (Mattisson *et al.* 2000). Use of Fick's second law gave the diffusion coefficient varying with pH and ionic strength of the gel, while the general model gave the diffusion coefficient independent of pH.

Because the diffusion of the incorporated antimicrobial is expected to be faster as the storage temperature increases, the antimicrobial will be released more quickly when coated foods are temperature abused. Since most food pathogens grow more quickly at higher temperatures up to their optimum growth temperatures, the more rapid release of the antimicrobials at higher temperatures can help maintain microbial safety of the coated foods.

The use of the mathematical models and the determined diffusion parameters, including both *D* and the partition coefficient between film/coating and food (*K*), will allow prediction of the time during which the antimicrobial remains above a critical inhibiting concentration in an edible film or coating used to inhibit microorganisms of concern (e.g., *L. monocytogenes*) in a target food. The models and the parameters can also be used to predict the optimum film/coating thickness and initial concentration of the antimicrobial for incorporation into the film or coating to obtain a desired antimicrobial activity against the targeted microorganisms in a selected food for a desired time.

Antimicrobial Coating for Nonthermally Processed Foods

Nonthermal food-processing technologies comprise an important area of study and application in food science and engineering. These technologies are being developed to satisfy consumer demand for fresh-like foods. They are intended to inhibit both spoilage and the growth of pathogenic microorganisms in foods without significant loss of flavor, color, taste, nutrients, viscosity, and functionality of the food by minimizing thermal effects on foods (Min and Zhang 2005). Fresh-like food products optimally processed by these technologies require appropriate packaging to preserve their qualities for desired shelf life during storage.

Antimicrobial edible films and coatings draw attention from the food and packaging industry because of increasing consumer demand for minimally processed products. Antimicrobial edible films and coatings can control microbial contamination occurring on the surface of the food during restorage after opening or because of package defects. In addition, since antimicrobial films and coatings have self-antimicrobial abilities, the need for chemical sanitization or sterilization of packaging materials may be obviated and aseptic packaging processes may be simplified (Hotchkiss 1997).

A combination use of nonthermal food processing and antimicrobial films and coatings is suggested because the bioactive films and coatings are expected to provide an additional barrier for the contamination and the growth of both spoilage and pathogenic microorganism in nonthermally processed food products. The benefit of nonthermal processing will not be altered because the antimicrobial films and coatings do not add either heat or synthetic chemicals to the nonthermally processed food. Both production of minimally processed foods and extension of their microbial stability can be made possible by combining the technologies.

Conclusion

The diffusion of antimicrobials in edible films and coatings is generally slower than that in foods. Thus, antimicrobial activity lasts longer when the antimicrobial compounds are incorporated in edible films and coatings compared with direct application on the surface of foods. Edible films and coatings can be designed to control the rate of antimicrobial diffusion to the film and food surfaces. Owing to their advantages, antimicrobial edible films and coatings have been intensively investigated for increasing microbial stability of foods during storage with safer and less expensive amounts of antimicrobials. Investigation of antimicrobial incorporation into edible films and coatings will continue in response to consumer demands for fresh-like foods. The combination use of edible coatings incorporating antimicrobials and nonthermal food processes is suggested for producing minimally processed and fresh-like food products with improved microbial stability.

References

Appendini P. & J. H. Hotchkiss. 1997. Immobilization of lysozyme on food contact polymers as potential antimicrobial films. *Packaging Technology and Science*. 10: 271–279.

Appendini P. & J. H. Hotchkiss. 2002. Review of antimicrobial food packaging. *Innovative Food Science & Emerging Technologies*. 3: 113–126.

Baldwin E. A. 1994. "Edible coatings for fresh fruits and vegetables: past, present, and future," in *Edible Coatings and Films to Improve Food Quality*, edited by J. M. Krochta, E. A. Baldwin & M. O. Nisperos-Carriedos. Lancaster, PA: Technomic, pp. 25–64.

Baldwin E. A., J. K. Burns, W. Kazokas, J. K. Brecht, R. D. Hagenmaier, R. J. Bender & E. Pesis. 1999. Effect of two edible coatings with different permeability characteristics on mango (*Mangifera indica* L.) ripening during storage. *Postharvest Biology and Technology*. 17: 215–226.

Biquet B. & T. P. Labuza. 1988. Evaluation of the moisture permeability of chocolate films as an edible moisture barrier. *Journal of Food Science*. 53: 989–998.

Bosch E. H., H. Van Doorne & S. De Vries. 2000. The lactoperoxidase system: the influence of iodide and the chemical and antimicrobial stability over the period of about 18 months. *Journal of Applied Microbiology*. 89: 215–224.

Brody A. L., E. R. Strupinsky & L. R. Kline. 2001. "Antimicrobial packaging," in *Active Packaging for Food Applications*, edited by A. L. Brody, E. R. Strupinsky & L. R. Kline. Lancaster, PA: Technomic, pp. 131–194.

Cagri A., Z. Ustunol & E. T. Ryser. 2004. Antimicrobial edible films and coatings. *Journal of Food Protection*. 67: 833–848.

Canillac N. & A. Mourey. 2001. Antibacterial activity of the essential oil of *Picea excelsa* on *Listeria*, *Staphylococcus aureus* and coliform bacteria. *Food Microbiology*. 18: 261–268.

Carson C. F. & T. V. Riley. 1995. Antimicrobial activity of the major components of the essential oil of *Melaleuca alternifolia*. *Journal of Applied Bacteriology*. 78: 264–269.

Cha D. S. 2002. Antimicrobial films based on Na-alginate and κ-carrageenan. *Lebens-mittel—Wissenschaft & Technologie*. 35: 715–719.

Cha D. S. & M. S. Chinnan. 2004. Biopolymer-based antimicrobial packaging: a review. *Critical Reviews in Food Science and Nutrition*. 44: 223–237.

Chen M.-J., Y.-M. Weng & W. Chen. 1999. Edible coating as preservative carriers to inhibit yeast on Taiwanese-style fruit preserves. *Journal of Food Safety*. 19: 89–96.

Coma V., I. Sebti, P. Pardon, A. Deschamps & H. Pichavant. 2001. Antimicrobial edible packaging based on cellulosic ethers, fatty acids, and nisin incorporation to inhibit *Listeria innocua* and *Staphylococcus aureus*. *Journal of Food Protection*. 64: 470–475.

Coma V., A. Martial-Gros, S. Garreau, A. Copinet, F. Salin & A. Deschamps. 2002. Edible antimicrobial films based on chitosan matrix. *Journal of Food Science*. 67: 1162–1169.

Conca K. R. & T. C. S. Yang. 1993. "Edible food barrier coatings," in *Biodegradable Polymers and Packaging*, edited by C. Ching, D. Kaplan & E. Thomass. Lancaster, PA: Technomic, pp. 357–369.

Cooksey K. 2000. "Utilization of antimicrobial packaging films for inhibition of selected microorganism," in *Food Packaging: Testing Methods and Applications*, edited by S. J. Rischs. Washington, DC: American Chemical Society, pp. 17–25.

Cosentino S., C. I. G. Tuberoso, B. Pisano, M. Satta, V. Mascia, E. Arzedi & F. Palmas. 1999. In-vitro antimicrobial activity and chemical composition of Sardinian *Thymus* essential oils. *Letters in Applied Microbiology*. 29: 130–135.

Cuppett S. L. 1994. "Edible coatings as carriers of food additives, fungicides and natural antagonists," in *Edible Coatings and Films to Improve Food Quality*, edited by J. M. Krochta, E. A. Baldwin & M. O. Nisperos-Carriedos. Lancaster, PA: Technomic, pp. 121–137.

Cuq B., N. Gontard & S. Guilbert. 1995. "Edible films and coatings as active layers," in *Active Food Packaging*, edited by M. L. Rooneys. New York: Blackie Academic & Professional, pp. 111–142.

Dawson P. L., G. D. Carl, J. C. Acton & I. Y. Han. 2002. Effect of lauric acid and nisin-impregnated soy-based films on the growth of *Listeria monocytogenes* on turkey bologna. *Poultry Science.* 81: 721–726.

Debeaufort F., J.-A. Quezada-Gallo & A. Voilley. 1998. Edible films and coatings: tomorrow's packagings: a review. *Critical Reviews in Food Science.* 38: 299-313.

Eswaranandam S., N. S. Hettiarachchy & M. G. Johnson. 2004. Antimicrobial activity of citric, lactic, malic, or tartaric acids and nisin-incorporated soy protein film against *Listeria monocytogenes, Escherichia coli* O157: H7, and *Salmonella gaminara. Journal of Food Science.* 69: FMS79–FMS84.

Franssen L. R., T. R. Rumsey & J. M. Krochta. 2004. Whey protein film composition effects on potassium sorbate and natamycin diffusion. *Journal of Food Science.* 69: C347–C350.

Gill A. O. & A. H. Richard. 2000. Surface application of lysozymes, nisin, and EDTA to inhibit spoilage and pathogenic bacteria on ham and bologna. *Journal of Food Protection.* 63: 1338–1346.

Goldberg S., R. Doyle & M. Rosenberg. 1990. Mechanism of enhancement of microbial cell hydrophobicity by cationic polymers. *Journal of Bacteriology.* 172: 5650–5654.

Halek G. W. & A. Garg. 1989. Fugal inhibition by a fungicide coupled to an ionomeric film. *Journal of Food Safety.* 9: 215–222.

Han J. H. 2000. Antimicrobial food packaging. *Food Technology.* 54: 56–65.

Han J. H. & J. D. Floros. 1997. Casting antimicrobial packaging films and measuring their physical properties and antimicrobial activity. *Journal of Plastic Film and Sheeting.* 13: 287–298.

Han J. H., J. M. Krochta, Y.-L. Hsieh & M. J. Kurth. 2000. Mechanism and characteristics of protein release from lactitol-based cross-linked hydrogel. *Journal of Agricultural and Food Chemistry.* 48: 5658–5665.

Hotchkiss J. H. 1997. Food-packaging interactions influencing quality and safety. *Food Additives and Contaminants.* 14: 601–607.

Kandemir N., A. Yemenicioglu, C. Mecitoglu, Z. S. Elmact, A. Arslanoglu, Y. Goksungur & T. Baysal. 2005. Production of antimicrobial films by incorporation of partially purified lysozyme into biodegradable films of crude exopolysaccharides obtained from *Aureobasidium pullulans* fermentation. *Food Technology and Biotechnology.* 43: 343–350.

Kester J. J. & O. R. Fennema. 1986. Edible films and coatings: a review. *Food Technology.* 40: 47–59.

Kim H.-W., K.-M. Kim, E.-J. Ko, S.-K. Lee, S.-D. Ha, K.-B. Song, S.-K. Park, K.-S. Kwon & D.-H. Bae. 2004. Development of antimicrobial edible film from defatted soybean meal fermented by *Bacillus subtilis. Journal of Microbiology and Biotechnology.* 14: 1303–1309.

Ko S., M. E. Janes, N. S. Hettiarachchy & M. G. Johnson. 2001. Physical and chemical properties of edible films containing nisin and their action against *Listeria monocytogenes. Journal of Food Science.* 66: 1006–1011.

Krochta J. M. & C. D. Mulder-Johnston. 1997. Edible and biodegradable polymer films. *Food Technology.* 51: 61–74.

Krochta J. M., S. Min & J. H. Han. 2005. Edible coatings containing bioactive agents. Oral symposium 56–52. Institute of Food Technologists Annual Meeting, New Orleans, LA. IFT.

Kussendrager K. D. & A. C. M. van Hooijdonk. 2000. Lactoperoxidase: physico-chemical properties, occurrence, mechanism of action and applications. *British Journal of Nutrition.* 84: S19–S25.

Kuu W. Y., R. W. Wood & T. J. Roseman. 1992. "Factors influencing the kinetics of solute release," in *Treatise on Controlled Drug Delivery*, edited by A. Kydonieuss, New York: Marcel Dekker, pp. 37–154.

Lis-Balchin M., M. Buchbauer, G. Ribisch & M. T. Wenger. 1998. Comparative antibacterial effects of novel *Pelargonium* essential oils and solvent extracts. *Letters in Applied Microbiology.* 27: 135–141.

Losso J. N., S. Nakai & E. A. Charter. 2000. "Lysozyme," in *Natural Food Antimicrobial Systems*, edited by A. S. Naidus. New York: CRC Press, pp. 185–210.

Mattisson C., P. Roger, B. Jonsson, A. Axelsson & G. Zacchi. 2000. Diffusion of lysozyme in gels and liquids: a general approach for the determination of diffusion coefficients using holographic laser interferometry. *Journal of Chromatography B.* 743: 151–167.

McHugh T. H. & J. M. Krochta. 1994. Sorbitol- vs glycerol-plasticized whey protein edible films: integrated oxygen permeability and tensile property evaluation. *Journal of Agricultural and Food Chemistry.* 42: 841–845.

Miller K. S. & J. M. Krochta. 1997. Oxygen and aroma barrier properties of edible films: a review. *Trends in Food Science & Technology.* 8: 228–237.

Min S. & J. M. Krochta. 2005. "Antimicrobial films and coatings for fresh fruit and vegetables," in *Improving the Safety of Fresh Fruit and Vegetables*, edited by W. Jongens. Abington, Cambridge, UK: Woodhead Publishing, pp. 454–492.

Min S. & Q. H. Zhang. 2005. "Packaging for non-thermal processing," in *Innovations in Food Packaging*, edited by J. H. Hans. London: Academic Press, pp. 482–500.

Min S., J. H. Han, L. J. Harris & J. M. Krochta. 2005. *Listeria monocytogenes* inhibition by whey protein films and coatings incorporating lysozyme. *Journal of Food Protection.* 68: 2317–2325.

Min S., L. J. Harris & J. M. Krochta. 2005a. Antimicrobial effects of lactoferrin, lysozyme, and the lactoperoxidase system and edible whey protein films incorporating the lactoperoxidase system against *Salmonella enterica* and *Escherichia coli* O157: H7. *Journal of Food Science.* 70: M332–M338.

Min S., L. J. Harris & J. M. Krochta. 2005b. Inhibition of *Salmonella enterica* and *Escherichia coli* O157: H7 on roasted turkey by edible whey protein coatings incorporating the lactoperoxidase system. *Journal of Food Protection.* 69: 784–793.

Min S., L. J. Harris & J. M. Krochta. 2005c. *Listeria monocytogenes* inhibition by whey protein films and coatings incorporating the lactoperoxidase system. *Journal of Food Science.* 70: M317–M324.

Ming X. W., H. George, J. W. Ayres & W. E. Sandine. 1997. Bacteriocins applied to food packaging materials to inhibit *Listeria monocytogenes* on meats. *Journal of Food Science.* 62: 413–415.

Nadarajah D., J. H. Han & R. A. Holley. 2005. Inactivation of *Escherichia coli* O157: H7 in packaged ground beef by allyl isothiocyanate. *International Journal of Food Microbiology.* 99: 269–279.

Natrajan N. & B. W. Sheldon. 2000. Inhibition of *Salmonella* on poultry skin using protein- and polysaccharide-based films containing a nisin formulation. *Journal of Food Protection.* 63: 1268–1272.

Orafidiya L. O., A. O. Oyedele, A. O. Shittu & A. A. Elujoba. 2001. The formulation of an effective topical antibacterial product containing *Ocimum gratissimum* leaf essential oil. *International Journal of Pharmaceutics*. 224: 177–183.

Ouattara B., R. E. Simard, G. Piette, A. Begin & R. A. Holley. 2000. Diffusion of acetic and propionic acids from chitosan-based antimicrobial packaging films. *Journal of Food Science*. 65: 768–773.

Ozdemir M. & J. D. Floros. 2003. Film composition effects on diffusion of potassium sorbate through whey protein films. *Journal of Food Science*. 68: 511–516.

Padgett T., I. Y. Han & P. L. Dawson. 1998. Incorporation of food-grade antimicrobial compounds into biodegradable packaging films. *Journal of Food Protection*. 61: 1330–1335.

Park H. J. 1999. Development of advanced edible coatings for fruits. *Trends in Food Science & Technology*. 10: 254–260.

Perez C., A. M. Agnese & J. L. Cabrera. 1999. The essential oil of *Senecio graveolens* (Compositae): chemical composition and antimicrobial activity tests. *Journal of Ethnopharmacology*. 66: 91–96.

Perez-Gago M. B. & J. M. Krochta. 2001. Denaturation time and temperature effects on oxygen permeability, film solubility and tensile properties of whey protein edible films. *Journal of Food Science*. 66: 705–710.

Pranoto Y., S. K. Rakshit & V. M. Salokhe. 2005. Enhancing antimicrobial activity of chitosan films by incorporating garlic oil, potassium sorbate and nisin. *Lebensmittel— Wissenschaft & Technologie*. 38: 859–865.

Pranoto Y., V. M. Salokhe & S. K. Rakshit. 2005. Physical and antibacterial properties of alginate-based edible film incorporated with garlic oil. *Food Research International*. 38: 267–272.

Reiter B. & G. Harnulv. 1984. Lactoperoxidase antibacterial system: natural occurrence, biological functions and practical applications. *Journal of Food Protection*. 47: 724–732.

Sebti I., F. Ham-Pichavant & V. Coma. 2002. Edible bioactive fatty acid-cellulosic derivative composites used in food-packaging applications. *Journal of Agricultural and Food Chemistry*. 50: 4290–4294.

Shah N. 2000. Effects of milk-derived bioactives: an overview. *The British Journal of Nutrition*. 84 (Suppl. 1): S3–S10.

Suppakul P., J. Miltz, K. Sonneveld & S. W. Bigger. 2003. Active packaging technologies with an emphasis on antimicrobial packaging and its applications. *Journal of Food Science*. 68: 408–420.

Weng Y.-M. & J. H. Hotchkiss. 1991. Headspace gas composition and chitin content as measures of *Rhizopus stolonifer* growth. *Journal of Food Science*. 56: 274–275.

Weng Y.-M. & J. H. Hotchkiss. 1992. Inhibition of surface molds on cheese by polyethylene film containing the antimycotic imazalil. *Journal of Food Protection*. 55: 367–369.

Wong D. W. S., W. M. Camirand & A. E. Pavlath. 1994. "Development of edible coatings for minimally processed fruits and vegetables," in *Edible Coatings and Films to Improve Food Quality*, edited by J. M. Krochta, E. A. Baldwin & M. O. Nisperos-Carriedos. Lancaster, PA: Technomic, pp. 65–88.

Yaman O. & L. Bayindirh. 2001. Effects of an edible coatings, fungicide and cold storage on microbial spoilage of cherries. *European Food Research and Technology*. 213: 53–55.

Zivanovic S., S. Chi & A. F. Draughon. 2005. Antimicrobial activity of chitosan films enriched with essential oils. *Journal of Food Science*. 70: M45–M51.

Chapter 4

BIO-MAP: MODIFIED-ATMOSPHERE PACKAGING WITH BIOLOGICAL CONTROL FOR SHELF-LIFE EXTENSION

James T. C. Yuan

Packaging is usually the last step in the food-manufacturing process, and it is essential for protecting foods. Modified-atmosphere packaging (MAP) has been applied successfully for the preservation of foods by extending the shelf life and improving the keeping quality. There are many useful resources on MAP and related technologies. Owing to consumers' requirement for fresh, close-to-fresh, or minimally processed foods (MPFs), and their strong demand for healthier, more tasty, and especially safer food, MAP has acquired a new dimension. This chapter provides a brief overview of the new MAP direction and its impact on shelf-life extension. MAP is explained in combination with other novel technologies such as biocontrol technology, high-pressure processing (HPP), and irradiation. Moreover, this chapter also demonstrates how the combined processes can be applied to enhance microbial inactivation efficacy while reducing the severity of individual treatment to maintain the quality of treated foods. Case studies are also shared.

Introduction

MAP has been researched and implemented in food preservation for years (Brody 1989; Farber 1989; Paine and Paine 1992; Mathlouthi 1994). Consumers enjoy fresh or near-fresh food products packaged in gas environment even after stored for many days. Food producers are benefiting from longer shelf life, which is translating into transport of their products for longer distance and better keeping quality for their products. Traditional MAP involves three key elements: packaging machine, packaging materials (such as pouch and tray), and gas mixture. Many useful reviews are readily available (Church 1994; Church and Parsons 1995). Commercial references are also easy to find at local market. However, traditional MAP has itslimits. For example:

1. MAP product has to be kept at refrigerated temperature or lower. Temperature is the key factor for a successful MAP, and nothing can replace the importance of a good temperature control because gas mixture is not a functional deliverable. Even these days, food producers have the wrong perception that as long as they provide good gas mixture and packaging materials, their product can withstand abusive temperature, and that is an absolutely wrong preconception. Gas mixture and/or packaging material are only good if storage temperature is well maintained.
2. MAP has no detrimental effect on the microbiological quality of packaged food products. Carbon dioxide (CO_2) may slow down or inhibit certain microbial growth with proper temperature management, but it would not eliminate the growth. Therefore, "garbage in, garbage out" rule applies well to MAP products.

Because of the recent demand for fresh foods, close-to-fresh foods, or MPFs, as well as the desire for healthier, tastier, and safer foods, MAP has acquired a new dimension. This chapter introduces a new concept of MAP which will not only preserve food products but also ensure the safety of MPFs.

New MAP Definition

In traditional MAP, the most commonly used gases are nitrogen (N_2), oxygen (O_2), carbon dioxide (CO_2), and argon (Ar), and their role is to preserve food by providing an inert environment so that all food constituents maximize their properties or characteristics during their shelf life. The new definition of MAP is modified-atmosphere processing, in which gas or gas mixture is used as a processing aid to actively deliver their functional features. It can also be called bio-MAP, as "bio" means "life," and that is what we would like to see—a new life of MAP. The packaging technology should shift to the front of combating unwanted microorganisms and should make a concerted effort to make food safe to consume. As mentioned above, gas or gas mixture is primarily used for inerting purpose, but in combination with other processing technologies they could serve a different molecular function. Laboratory case studies and illustrations are given for each of the following topics.

MAP with Biological Control

Biological control, also known as biocontrol, is the intentional introduction of parasites, predators, and/or pathogenic microorganisms to reduce or suppress populations of plant or animal pests; this has been the classic definition for centuries. In more recent decades, biocontrol use in the food industry refers to the application of nonpathogenic microorganisms to inhibit or retard the growth of pathogenic microorganisms. The concept of using biocontrol in MAPs has been studied for various food products (Holzapfel, Geisen, and Schillinger 1995; Adams and Nicolaides 1997; Amézquita and Brashears 2002), and these references would provide a good historical background on this combination.

To maximize the advantages of using biocontrol with MAP, lactic acid bacteria (LAB) in combination with CO_2-enriched atmosphere packaging may be used. The theory is that during good temperature storage, LAB would remain dormant, so it is transparent to consumers when cold chain integrity stays. However, if temperature abuse does occur, then LAB would rapidly initiate their growth themselves and outgrow pathogenic microorganisms (if present) and suppress or exclude them. LAB produce lactic acid, along with other substances such as bacteriocins and hydrogen peroxide, as a by-product of carbohydrate fermentation. This acid production lowers the pH of the food and inhibits the growth of many pH-sensitive organisms. Also, when LAB grow, lactic acid and other aromatic compounds are produced. Therefore, when consumers open a pack of chicken and if it smells like yogurt, they normally would not consume it, thereby ensuring food safety. To sum up, the key deliverable features of this technology are the following:

1. The selected LAB should be able to stay dormant under a good temperature-controlled environment. Otherwise, they would be considered spoilage microorganisms.
2. The selected LAB should be able to initiate themselves and outgrow in their food matrix very rapidly when cold chain breaks.

The criteria for a successful implementation of those features would be the following:

1. Selection of health-benefit microorganisms: One of the reasons why LAB are recommended in this technology is that LAB family is one of the most used groups in fermented foods, and consumers also benefit from their by-products (Gilliland 1986).
2. Optimum-atmosphere environment: LAB grow better under enriched CO_2 conditions. Hence, depending on the food products, balance gas could be either N_2 or Ar, with or without O_2.

As a case study, a laboratory trial was conducted to evaluate the growth of *Escherichia coli* in a ready-to-heat food product and to determine whether LAB could inhibit its growth when slight temperature abuse occurred. One set of samples was inoculated with *E. coli* alone and packaged. In another set, LAB plus *E. coli* were added. One LAB strain, LP, had good results in a previous experiment; hence, it was chosen for this evaluation. Commercially available rice bowl was chosen as the food medium and inoculated with a nonpathogenic strain of *E. coli* at 10^2 colony-forming units (CFU)/g. Samples were packaged under both a 50 : 50 mixture of N_2 and CO_2 and compressed air and incubated at three temperatures: 2°C, 7°C, and 20°C (room temperature). During the storage, the pH of the 2°C samples remained steady. At 7°C, by the seventh day of testing, the air samples had a very sour smell; especially the *E. coli* samples and the LP plus *E. coli* samples (with air in the package) had mold growth. The gas samples had only a very slight sour smell. This correlates with the microbial growth (Figure 4.1). A significant decrease in pH was seen for the LP plus *E. coli* in air environment between days 4 and 7 of the 7°C samples. The rest of the samples at that temperature had a large pH drop after 7 days of the study. The LP culture

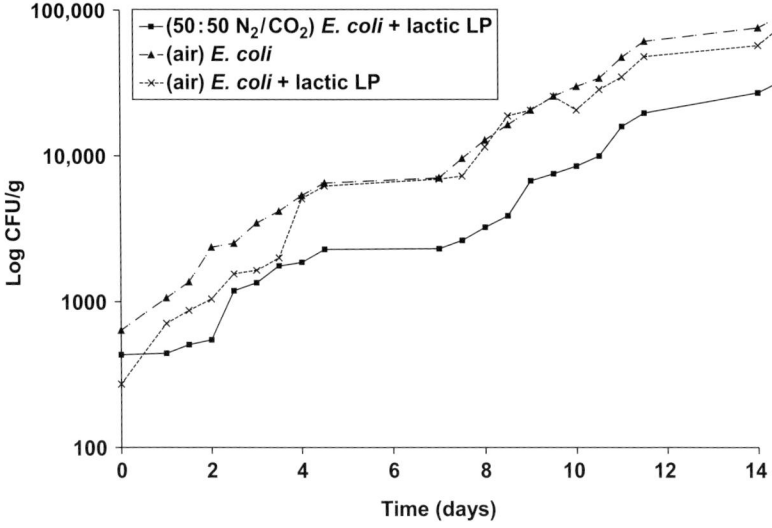

Figure 4.1. *Escherichia coli* plate count of chicken and rice bowl samples. Samples were inoculated with *E. coli* and lactic acid bacterium strain LP and packaged under two conditions. Shelf-life study was conducted at 7°C.

did not perform well in inhibiting *E. coli* growth under air environment. However, at room temperature, the inhibition produced by MAP plus LAB was evident in a $50:50$ N_2–CO_2 environment (Figure 4.2). Figure 4.3 illustrates the inhibition of *E. coli* probably due to a quick drop in pH. There was a relatively steady drop in pH of MAP-only or control samples compared with LAB samples, which had a sharp drop after 1.5 days of incubation at 20°C. Therefore, it appears that MAP in combination with LAB is able to slow down *E. coli* growth for a short period of time when temperature abuse is initiated.

MAP under Pressure

High-pressure treatment as a preservation technique for food has been known for over a century since Hite demonstrated in 1898 that microbial spoilage of milk could be delayed by application of pressure. Because of consumer demand for MPF-like products in recent years, there has been a renewed interest in this technology as an alternative to conventional thermal process. A range of pressure-treated products are on the Japanese markets for more than a decade, including fruit juices

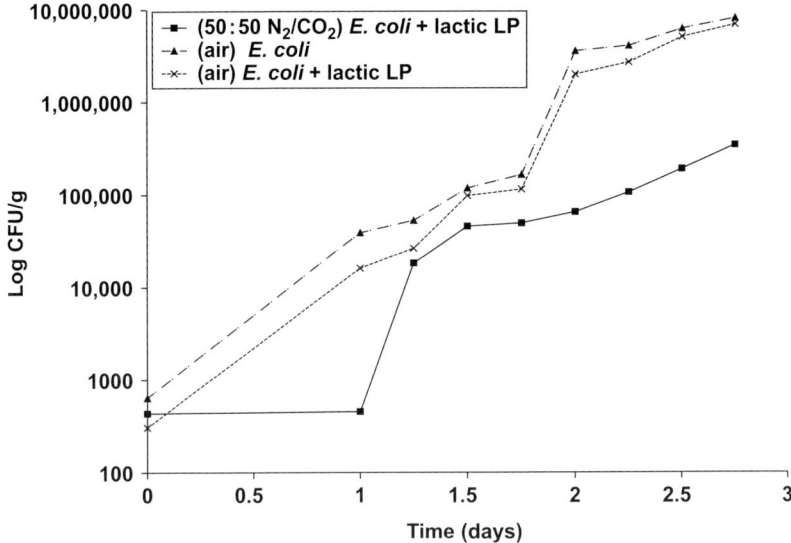

Figure 4.2. *Escherichia coli* plate count of chicken and rice bowl samples. Samples were inoculated with *E. coli* and lactic acid bacterium strain LP and packaged under two conditions. Shelf-life study was conducted at 20°C.

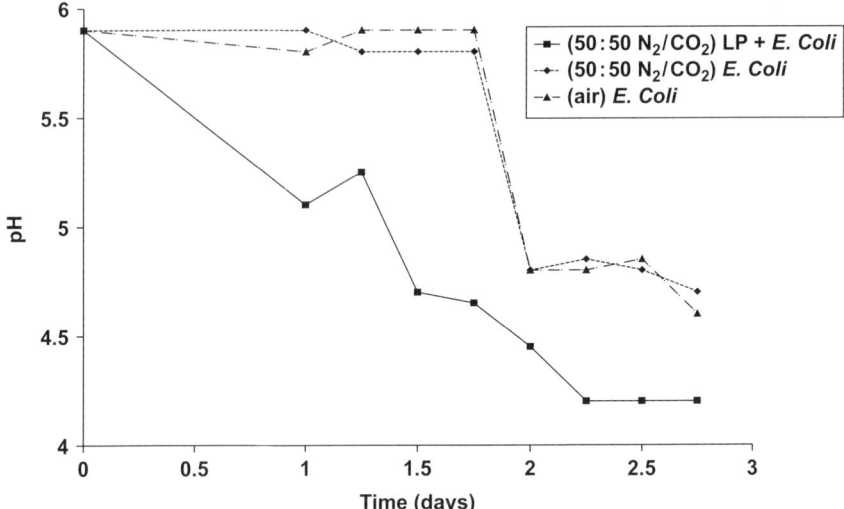

Figure 4.3. pH values of chicken and rice bowl samples. Samples were inoculated with *Escherichia coli* and lactic acid bacterium strain LP and packaged under two conditions. Shelf-life study was conducted at 20°C.

and rice cakes. In France, pressure-treated fruit juices are available. Recently, pressure-treated guacamole and oysters have become available on the U.S. markets.

Application of CO_2 in HPP has been evaluated mostly in liquid foods to improve the microbial inactivation efficacy of HPP (Hong, Park, and Pyun 1997, 1999; Karaman and Erkmen 2001). There was good synergism between both, with reductions in test microorganisms and enhancements in quality (Amanatidou *et al.* 2000), but further optimization needs to be studied before universal acceptance. The most popular and logical way of combining MAP with HPP is to inject gas mixture into each individual food container and treat it under pressure, therefore avoiding the possibility of contamination after repacking. Some potential advantages of MAP under pressure are the following:

1. improves the microbial reduction efficiency as a result of synergistic effect between HPP and gas mixture;
2. reduces the capital investment and operating cost of HPP because if the microbial reduction efficiency can be improved by incorporating gas mixture, then to achieve the same level of microbial reduction, HPP operating parameters would be reduced; therefore, HPP equipment and operating cost saving can be attained.

The effect of different gas atmosphere on the efficacy of HPP with regard to microbial inactivation is explained using *E. coli* and *Lactobacillus plantarum*. Also, a low temperature (40°C or less) was applied to isolate thermal impact. *Escherichia coli* was more pressure sensitive than *L. plantarum* as *L. plantarum* needed higher-pressure treatment at the same temperature and time conditions. At a process temperature of 10°C, CO_2 and O_2 atmosphere enhanced the inactivation of *E. coli* compared with other gases and the air control. O_2 was surprisingly much more effective than CO_2 (Figure 4.4). It probably can be explained by the toxic effect of high O_2 (Day 1996). At a process temperature of 40°C, CO_2 atmosphere enhanced the inactivation of *L. plantarum* compared with other gases and vacuum (the sample without any gas), whereas at a process temperature of 20°C, addition of gas had little or no effect (Figure 4.5). As differing from other gases, CO_2 atmosphere enhanced the inactivation of L. plantarum even at 20°C. There is a critical temperature and pressure condition above which all gases exist as fluid. The fluid above this point is said to possess both gas

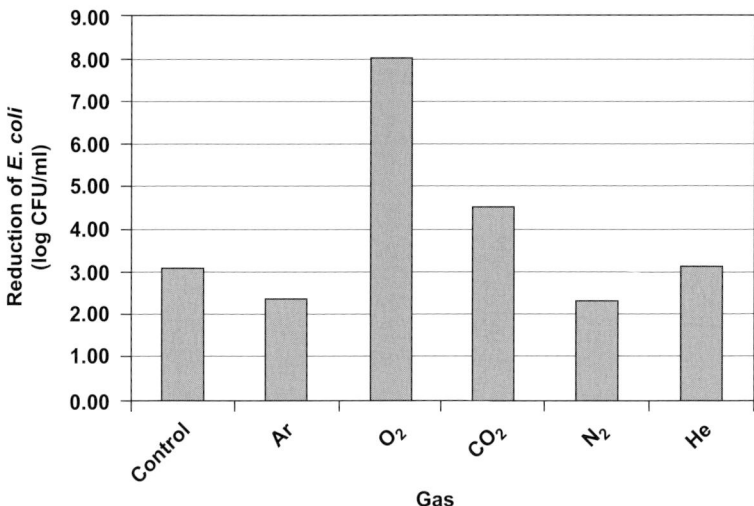

Figure 4.4. Inactivation of *Escherichia coli* by high-pressure processing (HPP) at 70 kpsi for 5 min under different atmospheres. The process temperature was kept at 10°C.

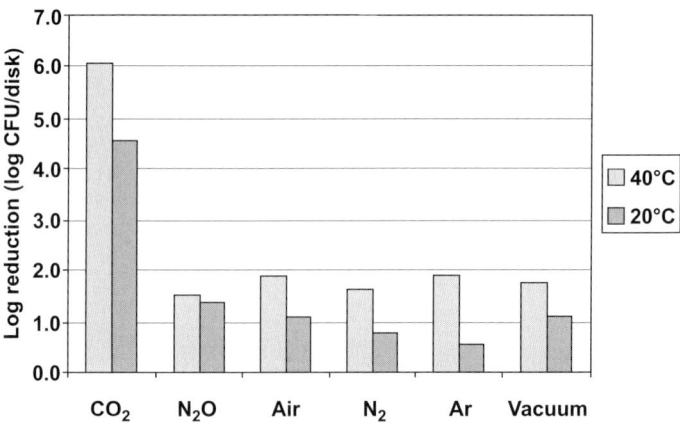

Figure 4.5. Inactivation of *Lactobacillus plantarum* by high-pressure processing (HPP) at 45 kpsi for 10 min under different atmospheres. The process temperature was kept at 20°C and 40°C.

(flow) and liquid (solubilization) properties. Fluids near critical region have some very interesting properties such as high heat capacity and solubility because of large-scale fluctuations around the region (Schieber 2003). The critical temperature and pressure conditions for different gases are presented in Table 4.1. Increased kill rate observed

Table 4.1. The critical temperature and pressure conditions for different gasses.

Compounds	Critical Pressure (MPa/psi)	Critical Temperature (°C)
Oxygen	5.08/736.8	−119
Nitrogen	3.40/493.1	−147
Carbon dioxide	7.38/1070.4	31
Hydrogen	1.30/188.5	−240
Nitrous oxide	7.27/1054.4	37
Helium	0.23/33.0	−267.95
Argon	4.90/710.4	−272.15

at 40°C with CO_2 addition is mainly due to the supercritical stage of CO_2 since inactivation without any gas (vacuum) did not show difference when the process temperature was changed from 40°C to 20°C. The mechanism of inactivation may be due to its acidity and enhanced diffusivity and solubility at supercritical stage, which was close to the critical point. Other gases (O_2, N_2, and Ar) are far from the supercritical region under the conditions tested. At this stage, they may clump and become bulky and less homogeneous than their free gas form, which may reduce their solubility and diffusivity characteristics. As the gases compress, they eventually start to cluster at some point, and the clusters become larger and larger. This molecular clustering is favored by high pressure and low temperature (Williams and Sprenger 1999).

Irradiated MAP

Some alternative technologies are being studied with MAP, including gamma irradiation (Grant and Patterson 1991a,b; Thayer and Boyd 2000). Irradiation is very effective on microbial reduction in all three forms. The role of gas atmosphere for irradiation is more of quality enhancement. During the irradiation process, if under normal air surrounding, it reacts with O_2 and creates ozone and other free radicals that could oxidize food and introduce off-odor and discoloration. Therefore, inert gas is needed in the package to protect irradiated food from being oxidized. The following laboratory experiment demonstrates the impact of inert gas atmosphere on irradiation-treated food products. The study used a three-strain mixture of *E. coli* and *Salmonella* sp. separately for the microbiological and quality analyses of inoculated ground beef after 17 days of storage at 4°C following electron beam irradiation and combined MAP. Packaging film was dual-layer type, which consisted of a barrier (outer layer) and permeable (inside layer). Dual-layer film is primarily used for meat products, and the idea is to isolate O_2 (from air) during irradiation process and through transportation and temporary storage to prevent products from oxidation and aerobic microbial spoilage. However, meat needs O_2 to display nice blooming color; hence, the second layer of permeable film is designed to let O_2 from air permeate into the package. When the product is ready to be placed on the shelf, the outer barrier layer is peeled and the second layer exposed to air. Since the second layer of film is permeable, it will let O_2 in from air and bloom the meat color. Microbial reductions for *E. coli* inoculated to 7 $\log_{10}$ CFU/g in ground beef irradiated to 1.2 kGy, 1.6 kGy, and 2.0 kGy were 2.0 $\log_{10}$ CFU/g, 2.7 $\log_{10}$ CFU/g, and 3.9 $\log_{10}$ CFU/g, respectively. The observed electron-beam-irradiation-induced microbial reductions for *E. coli* could be maintained without the growth of surviving cells on ground beef when combined with modified atmospheres of 100% N_2 (Figure 4.6) and 30% CO_2: 70% N_2 (Figure 4.7) followed by refrigerated storage (4°C) for up to 17 days. *Salmonella* exhibits the same pattern under 100% N_2 (Figure 4.8) and 30% CO_2 : 70% N_2 (Figure 4.9). Declines in pathogen viabilities were apparent when irradiation doses exceeded 1.6 kGy.

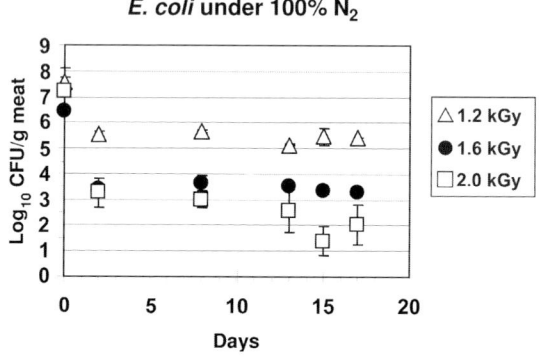

Figure 4.6. The viability of an *Escherichia coli* three-strain cocktail mixed with 80% fat-free ground beef, packaged using modified-atmosphere packaging (MAP), and electron beam irradiated followed by refrigerated storage (4°C) over time.

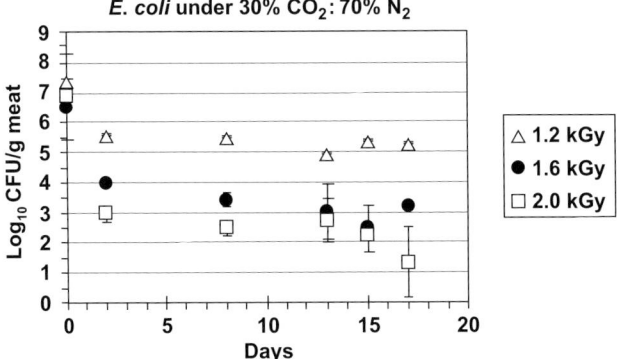

Figure 4.7. The viability of an *Escherichia coli* three-strain cocktail mixed with 80% fat-free ground beef, packaged using modified-atmosphere packaging (MAP), and electron beam irradiated followed by refrigerated storage (4°C) over time.

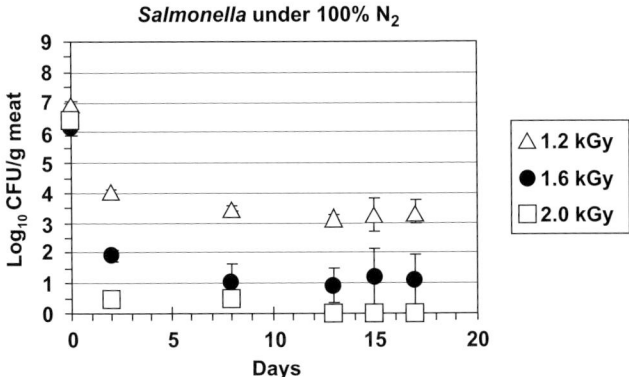

Figure 4.8. The viability of *Salmonella* sp. with 80% fat-free ground beef, packaged using modified-atmosphere packaging (MAP), and electron beam irradiated followed by refrigerated storage (4°C) over time.

With higher dosage (above 1.6 kGy), greater lipid oxidation occurred. Digitally recorded photographs of ground beef color during the refrigerated storage of irradiated and MAP beef revealed noticeable differences in color between MAP and the control sample. Microbial reductions were significant, and ground beef quality during storage could be increased using MAP combinations.

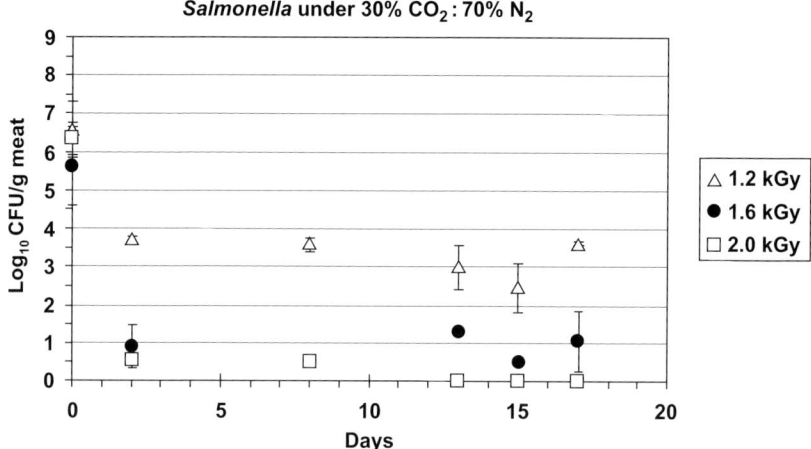

Figure 4.9.   The viability of *Salmonella* sp. with 80% fat-free ground beef, packaged using modified-atmosphere packaging (MAP), and electron beam irradiated followed by refrigerated storage (4°C) over time.

Conclusion

Changing for differences and especially for safer and better quality of foods is the way of ensuring our brands stay on the shelf. Packaging world also has to change or evolve into a new dimension, taking the lead of some new developments in response to food safety and security. Packaging gases have been with us for centuries, and we have just started to learn how to apply them to packages. Even though there not too many gases that we can use for food applications, and we probably exhausted all the possible gases and gases/packaging material combinations with different foods at different conditions. However, with novel processing technologies continuing to emerge, bio-MAP can certainly emerge itself into a new era, and food manufacturers will appreciate the diligence in making their products better and safer.

References

Adams, M.R. and Nicolaides, L. 1997. Review of the sensitivity of different foodborne pathogens to fermentation. *Food Control* 8(5/6): 227–239.

Amanatidou, A., Schlüter, O., Lemkau, K., Gorris, L.G.M., Smid, E.J., and Knorr, D. 2000. Effect of combined application of high pressure treatment and modified atmospheres on the shelf life of fresh Atlantic salmon. *Innovative Food Science & Emerging Technologies* 1: 87–98.

Amézquita, A. and Brashears, M.M. 2002. Competitive inhibition of *Listeria monocytogenes* in ready-to-eat meat products by lactic acid bacteria. *Journal of Food Protection* 65(2): 316–325.

Brody, A.L. 1989. *Controlled/Modified Atmosphere/Vacuum Packaging of Foods.* Food & Nutrition Press, Trumbull, CT.

Church, N. 1994. Development in modified-atmosphere packaging and related technologies. *Trends in Food Science & Technology* 5: 345–352.

Church, I.J. and Parsons, A.L. 1995. Modified atmosphere packaging technology: A review. *Journal of Science and Food Agricultural* 67: 143–152.

Day, B.P.F. 1996. Novel MAP for fresh prepared produce. *The European Food and Drink Review* Spring: 73–80.

Farber, J.M. 1989. Microbiological aspects of modified-atmosphere packaging technology— A review. *Journal of Food Protection* 54(1): 58–70.

Gilliland, S.E. 1986. *Bacterial Starter Cultures for Foods.* CRC Press, Boca Raton, FL.

Grant, I.R. and Patterson, M.F. 1991a. Effect of irradiation and modified atmosphere packaging on the microbiological safety of minced pork stored under temperature abuse conditions. *International Journal of Food Science & Technology* 26(5): 521–533.

Grant, I.R. and Patterson, M.F. 1991b. Effect of irradiation and modified atmosphere packaging on the microbiological and sensory quality of pork stored at refrigeration temperatures. *International Journal of Food Science & Technology* 26(5): 507–519.

Holzapfel, W.H., Geisen, R., and Schillinger, U. 1995. Biological preservation of foods with reference to protective cultures, bacteriocins and food-grade enzymes. *International Journal of Food Microbiology* 24: 343–362.

Hong, S., Park, W., and Pyun, Y. 1997. Inactivation of *Lactobacillus* sp. from kimchi by high pressure carbon dioxide. *Lebensmittel-Wissenschaft und-Technologie* 30: 681–685.

Hong, S., Park, W., and Pyun, Y. 1999. Non-thermal inactivation of *Lactobacillus plantarum* as influenced by pressure and temperature of pressurized carbon dioxide. *International Journal of Food Science & Technology* 34: 125–130.

Karaman, H. and Erkmen, O. 2001. High carbon dioxide pressure inactivation kinetics of *Escherichia coli* in broth. *Food Microbiology* 18: 11–16.

Mathlouthi, M. 1994. *Food Packaging and Preservation.* Blakie Academic & Professional, Glasgow, UK.

Paine, F.A. and Paine, H.Y. 1992. *A Handbook of Food Packaging.* 2nd ed. Blakie Academic & Professional, Glasgow, UK.

Schieber, J.D. 2003. Thermodynamics. Class notes, Fall, Illinois Institute of Technology, Chicago, IL.

Thayer, D.W. and Boyd, G. 2000. Reduction of normal flora by irradiation and its effect on the ability of *Listeria monocytogenes* to multiply on ground turkey stored at 7 deg. C when packaged under a modified atmosphere. *Journal of Food Protection* 63(12): 1702–1706.

Williams, J.M. and Sprenger, G.H. 1999. The 4th state of matter: the delta state. *Physics* 2: 1–18.

Chapter 5

PACKAGING FOR HIGH-PRESSURE PROCESSING, IRRADIATION, AND PULSED ELECTRIC FIELD PROCESSING

Seacheol Min and Q. Howard Zhang

Introduction

High-pressure processing (HPP), irradiation, and pulsed electric fields (PEF) have been developed as nonthermal food-preservation methods, which process foods at temperatures below those used for thermal pasteurization. They inactivate both spoilage and pathogenic micro-organisms in foods without significant loss of flavor, color, taste, nutrients, viscosity, and functionality of the foods by minimizing thermal effects on foods (Min and Zhang 2005). Packaging is a key factor to the feasibility of nonthermal processing because it plays a role in preserving the maximum of the intrinsic qualities of nonthermally processed food. Among many packaging materials, plastics have been intensively studied and used for foods processed by high pressure, irradiation, and PEF.

This chapter briefly describes HPP, irradiation, and PEF and reviews the effects of HPP and irradiation on packaging materials and the packaging materials for foods processed by high pressure and PEF.

High-Pressure Processing (HPP)

Foods are exposed to high pressure (100–800 MPa) during HPP for a short period, typically ranging from a few seconds to several minutes (3–5 min), depending on a target food temperature. Process temperature during the pressure treatment can be specified from below 0°C to above 100°C (Caner, Hernandez, and Harte 2004). Foods are pressurized using a pressure-transmitting medium (e.g., water). In a typical HPP, the food product to be pressurized is packaged in a flexible container, such as pouches and plastic bottles, and is loaded into a high-pressure chamber filled with the pressure-transmitting medium. The hydraulic fluid in the chamber is pressurized using a pump, and this pressure is transmitted through the package into the food product (FDA/CFSAN 2005). The food product is uniformly treated by high hydrostatic pressure because the pressure is evenly distributed throughout the product during HPP (Caner, Hernandez, and Harte 2004). No edge or thickness effect takes place (Caner, Hernandez, and Pascall 2000). The industrial equipments used have been designed for discontinuous batch processing (capacity: from 10 to 500 L/h) of solid, viscous, and particulated foods and for semicontinuous bulk processing (capacity: from 1 to 4 ton/h) of liquid foods (Yuste *et al.* 2002).

High pressure causes deprotonation of charged groups and disruption of salt bridges and hydrophobic bonds in microbial cell membranes, which results in conformational changes and denaturation of proteins. These conformational changes and denaturation of proteins cause the death of microorganisms (FDA/CFSAN 2005). Generally, bacteria are more resistant to HPP than yeasts and molds. High-pressure treatment alone is often not sufficient for substantial reduction of viable spore counts (Devlieghere, Vermeiren, and Debevere 2004). A subsequent pressure treatment of the germinated/germinating spores was shown to be an effective means of reducing spore counts (Devlieghere, Vermeiren, and Debevere 2004). Another concern about HPP in the food industry is the occurrence of pressure-resistant vegetative bacteria after successive pressure treatments (e.g., mutants of *Escherichia coli* and *Listeria monocytogenes*) (Devlieghere, Vermeiren, and Debevere 2004).

Critical process factors include pressure, magnitude and duration of high-pressure treatment, decompression time, treatment temperature, product initial temperature, product pH, product composition, composition of foods, the type of pressurizing medium, the type and the number of microorganisms, type of food packaging and concured processing aids, and packaging material integrity (Hoover 1997; Barbosa-Canovas *et al.* 1998).

HPP can be used to process both liquid and solid foods. Foods with a high acid content are particularly good candidates for HPP technology.

High-pressure-processed products are commercially available in the United States, European, and Japanese retail markets. Examples of high-pressure-processed products commercially available in the United States are fruit smoothies, fruit jellies and jams, fruit juices, tomato salsa, guacamole, cooked ready-to-eat meats, oysters, ham, chicken strips, pourable salad dressing, and rice cakes (Caner, Hernandez, and Harte 2004; Devlieghere, Vermeiren, and Debevere 2004).

Packaging for HPP

The effects of HPP on packaging materials are important to be investigated because foods to be treated are generally packaged prior to high-pressure treatments (Ozen and Floros 2001). Reversible response of a whole package to compression is crucial to the success of the high-pressure treatment. Packaging materials for HPP are required to be flexible enough to withstand the compression forces while maintaining physical integrity. They must recover their initial volumes after the pressure is released (Caner, Hernandez, and Harte 2004). This is a reason why metal cans, glass bottles, and paperboard-based packages are not well suited for HPP (Lambert *et al.* 2000; Caner, Hernandez, and Harte 2004).

Due to the fact that air or gases are very compressible under high pressure, the more the headspace, the bigger the deformation strains on the packaging materials. The presence of headspace must be kept as small as possible (Lambert *et al.* 2000).

Various flexible systems including polypropylene (PP), polyester tubes, polyethylene (PE) pouches, and nylon cast PP pouches are currently used for HPP products. HPP has shown compatibility with the most common plastic packaging materials. However, a metal layer, such as an aluminum (Al) layer, in a multilayered packaging material negatively affected its barrier properties. This could be related to the promotion in rupture in a less elastic component, the metal layer (Caner, Hernandez, and Harte 2004). An inorganic coating such as aluminum trioxide (Al_2O_3) and silicon dioxide (SiO_2) in a multilayered packaging material positively affected its mechanical and barrier properties (Caner, Hernandez, and Harte 2004).

Failure in the integrity of the package after HPP is very important for food safety. Thus, the effects of HPP on the physical and barrier properties, delamination, and sealing integrity of packages have been studied by many researchers (Table 5.1).

Table 5.1. Effects of high-pressure processing (HPP) on the properties of packaging materials.

Property	Packaging Materials	HPP Condition	Effect	Reference
Barrier properties	PP/EVOH/PP, OPP/PVOH/PE, KOP/CPP, PET/Al/CPP	400–600 MPa	No significant changes in permeability of O_2 and water vapor	Masuda *et al.* (1992)
	PET/SiOx/polyurethane (PU) adhesive /LDPE, PET/Al$_2$O$_3$/PU/LDPE, PET/PVDC/nylon/HDPE/PE, PE/nylon/EVOH/PE,PE/ nylon/PE, PET/EVA, PP	600–800 MPa, 5, 10, 20 min, 45°C	No significant changes in the permeability of O_2, CO_2, and water vapor	Caner, Hernandez, and Pascall (2000)
	PA/PE, PA/PE, PET/PVDC/PE, PA/PP/PE	200, 350, 500 MPa, 30 min	No significant changes in the permeability of O_2 and water vapor	Lambert *et al.* (2000)
	PE coated with SiOx and PP	800 MPa, 2 min, 25°C	No significant O_2-barrier property change	Lambert *et al.* (2000)
	LLDPE/EVA/EVOH/EVA/ LLDPE,PET/Al/PP	400 MPa, 30 min, 60°C	No significant water vapor-barrier property change	Lambert *et al.* (2000)
	Metallized PET/EVA/LLDPE	600–800 MPa, 5, 10, 20 min, 45°C	Significant increases in the permeability of O_2, CO_2, and water vapor	Caner, Hernandez, and Pascall (2000)
Mechanical properties	PET/SiOx /LDPE,PET/Al2O3/ LDPE, PET/PVDC/nylon/HDPE/ PE,PE/nylon/EVOH/PE, PE/nylon/ PE,PP/nylon/PP, PET/EVA/PET	600–800 MPa, 5, 10, 20 min, 45°C	No significant changes in mechanical properties	Caner, Hernandez, and Harte (2004)

	Material	Conditions	Result	Reference
	LLDPE/EVA, EVOH/EVA/LLDPE, PET/Al/PP PA/PE, PET/PVDC/PE, PA/PE surlyn, PA/PP/PE	400 MPa, 30 min, 60°C 200–500 MPa, 30 min, 20°C	No significant changes in tensile strength and elongation Tensile strength increased (change <25%)	Caner, Hernandez, and Harte (2004) Lambert et al. (2000)
Volatile migration	PP, PE/nylon/EVOH/PE	800 MPa, 10 min, 60°C	No significant absorption of D-limonene	Caner et al. (2004)
	LDPE, EVA	400 MPa, 10 min	Decrease in the sorption of D-limonene	Masuda et al. (1992)
	LDPE/HDPE/LDPE	50 MPa, 23°C	Decrease in the permeation rate of p-cymene to LDPE/HDPE/LDPE	Goetz and Weisser (2002)
	PE/nylon/Al/PP pouches, nylon/EVOH/PE pouches	150, 300, 600 MPa, 30 s, 30, 50, 75°C	Decrease in the migration of 1, 2-propanediol (PG)	Schauwecker et al. (2002)
Delamination	PA/PE, PET/PVDC/PE, PA/PE surlyn, PA/PP/PE	200, 350, 500 MPa, 30 min	No delamination	Lambert et al. (2000)
	PA/PE	200, 350, 500 MPa, 30 min	Delamination	Lambert et al. (2000)
	PA/PE	500 MPa, 300 s	Delamination	Goetz and Weisser (2002)
	PE/nylon/Al/PP	200, 690 MPa, 10 min, 85, 90, 95°C	Delamination	Schauwecker et al. (2002)

Barrier Properties

Most of the tested materials were compatible with HPP at tested conditions (Table 5.1). About 12% is considered as an allowable deviation for the changes in oxygen- and water vapor-barrier properties after a high-pressure treatment (Lambert *et al.* 2000). Packaging materials constituted with ethylene vinyl alcohol (EVOH) and polyvinyl alcohol (PVOH) were considered compatible with high-pressure treatment (Masuda *et al.* 1992). Water vapor and oxygen permeabilities of several laminated plastic films (PP/EVOH/PP, oriented polypropylene (OPP)/ PVOH/PE, KOP/cast polypropylene (CPP), polyethylene terephthalate (PET)/Al/CPP) were not affected from high pressures between 400 and 600 MPa (Ozen and Floros 2001). Oxygen-barrier properties of PE coated with SiOx and PP materials were not changed by a treatment at 800 MPa (2 min, 25°C) (Lambert *et al.* 2000). Significant increases in the permeability of oxygen, carbon dioxide, and water vapor were observed in a metallized PET film after a high-pressure treatment (Caner, Hernandez, and Pascall 2000; Caner, Hernandez, and Harte 2004). Results showed that metallized PET was most severely affected by HPP, as its permeance values for oxygen, carbon dioxide, and water vapor increased as much as 150%.

Mechanical Properties

Mechanical properties of all the materials listed in Table 5.1 were not negatively affected by high-pressure treatments. Tensile strength of polyamide (PA)/PE, PET/polyvinylidene chloride (PVDC)/PE, PA/PE surlyn, and PA/PP/PE increased after a high-pressure treatment, indicating the packages became more rigid and less flexible, but this may not be practically different when 25% is considered as an allowable deviation that industries use (Lambert *et al.* 2000).

Migration of Flavor Compounds in Packaging Materials

During compression by high pressure, a free volume of a plastic matrix decreases, and it loses its capacity for absorbing flavor compounds from food (Caner, Hernandez, and Harte 2004). As the pressure is released, the plastic material ideally quickly recovers its original dimensions, and thus sorption and diffusion proceed as expected at normal atmospheric pressure (Caner, Hernandez, and Harte 2004). Some materials failed to regain their free volume, reducing the migration of flavor compounds in the materials (Table 5.1). Masuda *et al.* (1992) reported a decrease in the sorption of D-limonone into low-density polyethylene (LDPE) and ethylene vinyl acetate (EVA) films as a result of the treatment under 400 MPa pressure for 10 min. Kubel *et al.* (1996) investigated the effect of HPP on the sorption of aroma compounds, *p*-cymene, and acetophenone into plastic films. They found that the sorption of aroma compounds was lower in films exposed to 500 MPa pressure compared to nonpressurized films. Transition of the films to the glassy state at higher pressures was suggested as the reason for the decrease in the sorption of the aroma compounds (Kubel *et al.* 1996).

Delamination

Delamination during HPP can be caused by the nature of the glue used, presence of air pocket in packaging materials, and different compressibility of materials laminated (Lambert *et al.* 2000; Goetz and Weisser 2002). A PE/nylon/Al/PP laminate was delaminated after high-pressure treatments (Schauwecker *et al.* 2002) (Table 5.1). Delamination between the PP and the Al layers was observed in meal-ready-to eat (MRE) pouches treated at $\geqslant$ 200 MPa and 90°C for 10 min (Schauwecker *et al.* 2002). Al has a very low compressibility compared to the others, and this could result in delamination.

Sealing

Sealing is another critical point in producing high-pressurized food with flexible packages as described previously. Thus, the closure or seal integrity should be closely investigated because leakage may occur through sealed areas after HPP (Caner, Hernandez, and Harte 2004). However, reports on the effects of high-pressure treatment on sealing integrity are lacking. Lambert *et al.* (2000) reported that heat-seal strength of the multilayer plastic packages of PA/PE, PET/PVDC/PE, PA/PE surlyn, and PA/PP/PE was not modified by a high-pressure treatment (200–500 MPa, 20°C, 30 min). Sealability of a copolymer of ethylene and methacrylic acid increased after a HPP at 600 MPa for 60 min (Dobias *et al.* 2004).

Irradiation

Irradiation exposes food to a source of ionizing radiation sufficient to create positive and negative charges. Radiation sources approved for food used are gamma rays produced by the radioisotopes cobalt-60 or cesium-137, machine-generated X-rays [maximum energy of 5 million electron volts (MeV)], and electrons (maximum energy of 10 MeV) (Olson 1998). Depending on the dose of radiation energy applied, foods may be pasteurized to reduce or eliminate pathogens, or they may be sterilized to eliminate all microorganisms (Olson 1998).

Radiation inactivates microorganisms by damaging genetic materials (Grecz, Rowley, and Matsuyama 1983). This damage prevents multiplication and terminates most cell functions. Radiation also directly damages other components of microbial cells such as membranes, enzymes, and plasmids (Smith and Pillai 2004).

The amount of vitamin loss due to food irradiation is affected by several factors, including dose, temperature, presence of oxygen, and food type. Generally, radiation at low temperatures in the absence of oxygen reduces any vitamin loss in foods, and storage of irradiated foods at low temperatures also helps prevent future vitamin loss (Olson 1998).

Irradiation is being used for foods in more than 30 countries. The Joint FAO/IAEA/WHO Expert Committee confirmed that irradiation up to 10 kGy does not produce toxicological hazards and nutritional or microbiological problems in foods (Lee *et al.* 2004). The United States currently has the most widespread approvals for the use of irradiation for food (Ozen and Floros 2001). Dried foods are less sensitive to irradiation than hydrated ones, and their irradiation has been authorized at a maximum dose of 10 and 30 kGy in Korea and in the United States, respectively (Lee *et al.* 2004). Spices are the most commonly irradiated food (Olson 1998).

Packaging for Irradiation

Foods are generally prepackaged before irradiation to prevent recontamination. The use of irradiation is also becoming a common treatment to sterilize packages in aseptic processing of foods and pharmaceuticals (Ozen and Floros 2001). Therefore, the effects of irradiation on the packaging materials, including any coextruded and laminated materials, must be studied. Any packaging materials must be accepted by FDA before use in food irradiation because gases (e.g., hydrogen) and low-molecular-weight hydrocarbons and halogenated polymers formed during irradiation at doses accepted for food use have a potential to migrate into foods (Kilcast 1990; Lee *et al.* 1996; Olson 1998). Free radicals can be formed from plastics by irradiation, which can be incorporated in plastics in crystalline regions and age the plastics (Buchalla, Schuttler, and Bogl 1993; Ozen and Floros 2001). Marque *et al.* (1995) detected alkyl radicals, which were oxidized to peroxyl radicals in the presence of air after ionization treatment of PP at 40 kGy. Any toxic substances should not be transmitted from packaging materials to foods (Barbosa-Canovas *et al.* 1998). Materials for irradiation are listed in 21 CFR 179.45.

Some chemical and physical properties of polymeric packaging materials can be changed by irradiation (Ozen and Floros 2001). The changes depend on the type of polymer, processing exposure, and irradiation conditions (Crook and Boylston 2004). Predominant reactions during irradiation in most plastics used for food packaging [e.g., PE, PP, polystyrene (PS)] are cross-linking and chain scission (Ozen and Floros 2001). Cross-linking can decrease elongation, crystallinity, and solubility and increase the mechanical strength of the plastics. Chain scission can decrease the chain length of plastic materials, providing free volume in the plastics, and can produce hydrogen, methane, and hydrogen chloride for chlorine-containing polymers under vacuum (Ozen and Floros 2001). In the presence of oxygen, additional chain scissions would be able to form peroxide, alcohol, and various low-molecular-weight oxygen-containing compounds (Ozen and Floros 2001).

Effects on Barrier Properties

Tested barrier properties of LDPE, high-density polyethylene (HDPE), PET, EVA, PS, and biaxially oriented polypropylene (BOPP) were not significantly changed by irradiation (Table 5.2). The oxygen-barrier capability of the LDPE bags was improved by 7.7 and 4.5% after irradiation at 1.5 and 3.2 kGy, respectively. Irradiation did not affect water vapor permeability nor did it affect the stiffness of LDPE films (Han *et al.* 2004). Radiation doses of 5, 10, and 30 kGy did not cause any statistically significant changes in the gas permeability (O_2 and CO_2) and water vapor permeability of EVA, HDPE, PS, BOPP, LDPE, and

Table 5.2. Effects of irradiation on the properties of packaging materials.

Property	Packaging Materials	HPP Condition	Effect	Reference
Barrier properties	LDPE, HDPE, PET, PVC	8 kGy	No significant changes in gas permeability	Crook and Boylston (2004)
	PE pouch	25 kGy	No significant changes in the permeability of O_2 and water vapor	Rojas De Gante and Pascat (1990)
	EVA, HDPE, PS, BOPP, LDPE	5, 10, 30 kGy (γ-radiation)	No significant changes in the permeability of CO_2, O_2, and water vapor	Goulas et al. (2002)
Mechanical properties	PS	100 kGy (γ-irradiation)	No changes	Pentimalli et al. (2000)
	PP, LDPE polylactic acid	0.5–2.0 kGy (^{60}Co irradiation)	No changes	Krishna murthy et al. (2004)
	EVA, HDPE, PS, BOPP, LDPE	5, 10 kGy	No significant change in tensile strength, % elongation at break, Young's modulus	Goulas et al. (2002)
	HDPE, BOPP	30 kGy	Tensile strength decreased	Goulas et al. (2002)
	LDPE	30 kGy	% elongation at break decreased	Goulas et al. (2002)
Volatile formation	LDPE, OPP	<25 kGy	H_2O_2, carbonyl compounds (ketones and aldehydes)	Rojas De Gante and Pascat (1990)
	PP	40 kGy	Alkyl radicals	Marque et al. (1995)
	PET	25 kGy (cesium 137)	Formic acid, acetic acid, 1, 3-dioxolane, and 2-methyl-1, 3 dioxolane	Komolprasert et al. (2001)

ionomer (Goulas *et al.* 2002). Effects of irradiation on the characteristics of diffusion and the sorption of octan, ethyl hexanoate, and D-limonene were investigated (Matsui *et al.* 1991). With increasing radiation dose, the diffusion coefficient through electron beam-radiated EVA film increased. The coefficient increase was thought to be related to chain scissions in EVA film caused by irradiation.

Effects on Mechanical Properties

Mechanical properties of PS were not significantly changed by a 100 kGy irradiation (Table 5.2). Mechanical properties of HDPE, BOPP, and LDPE did not change at a low dose but changed at a high dose of irradiation (Table 5.2). The effect of gamma-radiation doses (5, 10, and 30 kGy) on monolayer flexible packaging films, EVA, HDPE, PS, BOPP, and LDPE, was studied (Goulas *et al.* 2002). The results showed that the tensile strength, percentage elongation at break, and Young's modulus of all the films remained unaffected after absorbed doses of 5 and 10 kGy. The tensile strength of HDPE and BOPP films irradiated at a dose of 30 kGy decreased. The percentage elongation at break of LDPE films irradiated at 30 kGy decreased. Young's modulus of all the tested films remained unchanged (Goulas *et al.* 2002).

Effects on Volatile Formation

Generally speaking, volatiles are formed from plastic materials by irradiation. An irradiation formed 63 different volatiles from PE, PET, and OPP (El Makhzoumi 1994). Rojas De Gante and Pascat (1990) reported that hydroperoxides and carbonyl compounds such as ketones and aldehydes formed by an irradiation of LDPE and OPP (<25 kGy) (Table 5.2.). Major volatile compounds that evolved from the PET specimens are formic acid, acetic acid, 1, 3-dioxolane, and 2-methyl-1, 3 dioxolane (Komolprasert *et al.* 2001). A 25 kGy irradiation (caesium[137]) had a significant effect on increasing the volatile compounds in crystalline and oriented semirigid PET homopolymer (Komolprasert *et al.* 2001). Irradiation on LDPE, HDPE, PET, and polyvinyl chloride (PVC) packaging materials can release hydrogen, carbon dioxide, carbon monoxide, and methane gases and form volatile oxidation products including peroxides, alcohols, aldehydes, ketones, and carboxylic acids (Crook and Boylston 2004).

Irradiation and Modified Atmosphere Packaging

Modified atmosphere packaging (MAP) has been intensively investigated to minimize quality changes or even to improve the quality of irradiated foods during storage. Lipid oxidation in ground beef ionized at 2.0 kGy and packaged aerobically progressed at a significantly higher rate compared to those irradiated and packaged under vacuum or nonirradiated control beef (Murano, Murano, and Olson 1998). This was considered due to the production of hydroperoxyl radicals by the reaction of radiolytic products formed during irradiation with oxygen. The radicals can initiate lipid oxidation. Thus, reducing or eliminating oxygen from the packaging environment will reduce the radical formation and lipid oxidation (Kusmider *et al.* 2002). A carbon monoxide (<1%) packaging reduced lipid oxidation of ground beef treated at 4.5 kGy and provided a very stable, cherry-red product color, as indicated by instrumental and sensory analysis (Kusmider *et al.* 2002). Ahn *et al.* (2003) reported positive effects of combination of irradiation and MAP [CO_2 (100%) or CO_2/N_2 (25%/75%)] in reducing the loss of the red color of irradiated (5 and 10 kGy) sausage. Nitrosamine contents in meat products can be reduced by irradiation during storage, and the reduction can be more extensive with MAP than with aerobic packaging (Song *et al.* 2003).

However, an attention needs to be drawn in applying MAP for the packaging of nonthermally processed foods since MAP can selectively change the microbiology of foods. The suppression of aerobic spoilage microorganisms will decrease competition for growth and provide sufficient time for the growth of pathogenic microorganisms, especially if the MAP foods received a nonsterilizing treatment (Hotchkiss and Banco 1992; Marth 1998).

Pulsed Electric Field (PEF) Processing

PEF treatment uses high-intensity electric field pulses generated between two electrodes. Nonthermal treatment is attained by minimizing temperature increase in treatment zones using a very short pulse width of treatment time (i.e., microseconds).

A commercial PEF-processing system is shown in Figure 5.1. The system comprises a high-voltage pulse generator, a PEF treatment chamber, a data-logging system, and an aseptic drink processor (fluid-handling system). The pulse generator, connected to the treatment

Figure 5.1. A commercial scale pulsed electric field (PEF) processing system (OSU 6, The Ohio State University, OH, USA).

79

chambers, generates electric pulses. PEF are delivered to foods inside the PEF treatment chamber. A commercial scale PEF system located at the Ohio State University was tested for a cost evaluation. The total operation cost consisted of three parts: equipment, personal, and utilities. Utilities included water, electricity, and steam with local industrial prices. The cost for packaging and food materials was not included. Production rate used for the evaluation was 1000 L/h. The evaluation-reported total operational cost was 6 to 7 cents per liter for PEF-treated orange and tomato juices, which was considered competitive with a conventional thermal processing of the juices (Jin and Zhang 2002).

PEF inhibits microorganisms by damaging microbial cell membranes (Devlieghere, Vermeiren, and Debevere 2004). This damage leads to ion leakage, metabolite losses, protein releases, and increased uptakes of drugs, molecular probes, and DNA (Kinosita and Tsong 1977; Benz and Zimmermann 1980).

The factors determining the efficiency of the microbial inactivation by PEF can be classified based on treatment parameters, product parameters, and microbial characteristics. The main treatment parameters that affect microbial inactivation by PEF are electric field strength, PEF treatment time, pulse width, pulse shape, and treatment temperature (Hulsheger and Niemann 1980; Zhang *et al.* 1994, 1995; Zhang, Barbosa-Canovas, and Swanson 1995; Barbosa-Canovas *et al.* 1999). The most critical product parameters include electric conductivity, density, viscosity, pH, and water activity (Zhang, Barbosa-Canovas, and Swanson, 1995; Vega-Mercado *et al.* 1996; Wouters, Alvarez, and Raso 2001; Min, Reina, and Zhang 2002). The type of microorganism, microbial cell size or shape, growth stage of microorganisms, and microbial stress adaptation have been reported to influence the efficiency of PEF inactivation (Grahl and Markl 1996; Raso *et al.* 1998; Wouters, Bos, and Ueckert 2001; Evrendilek and Zhang 2003). Various types of juices, milk, yogurt, and liquid egg have been tested by PEF, and its benefits for their processes have been revealed (Yeom *et al.* 2002).

Packaging for PEF-Processed Food Products

Initially, high quality of PEF-treated food products can be extended over time by selecting proper juice packaging materials and methods. Plastic packaging materials have been tested for packaging of PEF-processed foods. The shelf life of foods packaged into plastics depends on the permeation of gas and water vapor through packages because a significant amount of food deterioration results from oxidation and changes in the water content. This is also the case for PEF-processed foods. Thus, the permeation values are very important in determining packaging materials for PEF-processed food products. Aseptic food packaging is considered the most appropriate way of packaging for PEF-processed food products (Qin *et al.* 1995). Plastics and paper-laminated materials are widely used as packaging materials for aseptic food packaging. These are usually thermoformed to produce food containers during aseptic packaging.

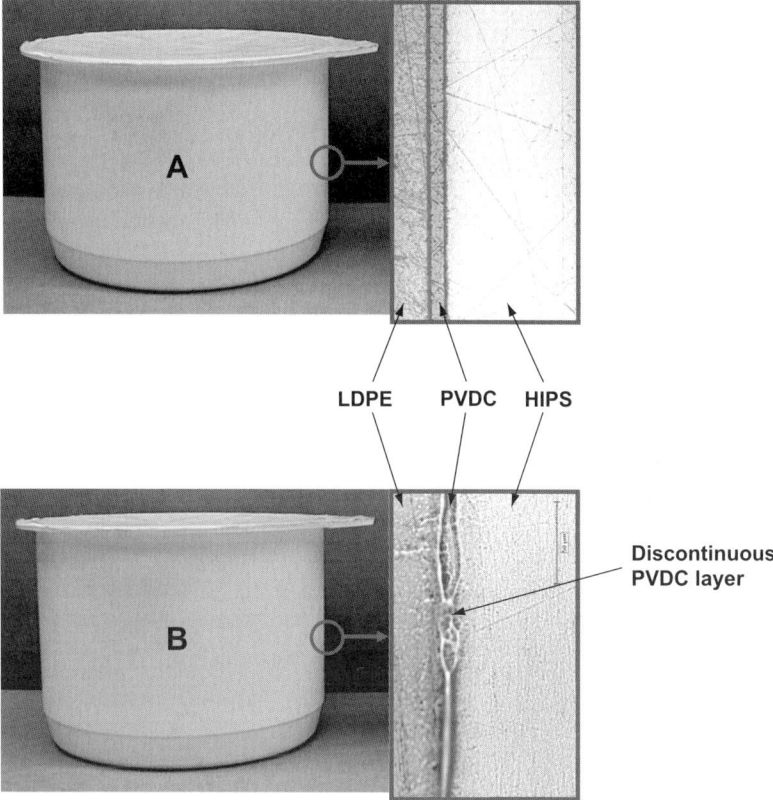

Figure 5.2. Thermo-formed containers and the microscopic images of their side walls. A LDPE/PVDC/HIPS multilayered material was thermoformed to form the containers. Containers A and B were produced at optimized and nonoptimized thermoforming conditions, respectively. LDPE, low-density polyethylene; PVDC, polyvinylidene chloride.

When laminated materials are used, delamination or destructions of barrier layers (e.g., PVDC) during thermoforming must be avoided because those will negatively affect barrier and mechanical properties of thermoformed containers (Figure 5.2). For aseptic packaging of PEF-processed food products, thermoforming conditions should be optimized so that the material used does not lose its barrier or mechanical properties.

The effects of packaging materials on the quality preservation of orange juice treated by PEF (35 kV/cm, 59 µs, pilot plant scale PEF system) were studied (Ayhan *et al.* 2001). Some chemical and physical properties of PEF-treated orange juice packaged in four different packaging materials, sanitized glass, polyethylene terephthalate (PET), HDPE, and LDPE bottles, were evaluated. Glass bottles and PET bottles were effective at retarding degradations of flavor compounds, vitamin C, and color of PEF-treated orange juice during storage at 4°C for 112 days. Vitamin C and the flavor compounds were labile in polyethylene (HDPE and LDPE) bottles, which might be due to their relatively low barrier property of polyethylene to oxygen.

The degradation of lycopene of PEF-processed tomato juice in a PP tube was found most significant during the first 7 days of the storage at 4°C (Min, Jin, and Zhang 2003). The main cause of carotenoid degradation in foods is oxidation (Thakur, Singh, and Nelson 1996), and thus the significant reduction was considered due to oxygen availability in the headspace of the PP tubes (Min, Jin, and Zhang 2003). The MAP, which limits oxygen in the headspace, may be applied as a complement to PEF to reduce oxidation of PEF-processed food products.

Conclusions

Nonthermal processing has been developed to satisfy consumers' demand for fresh-like foods and, at the same time, to provide food microbial safety. The benefit of nonthermal processing will be realized when nonthermally processed foods maintain their initially high quality for extended storage time. The extension of the quality will be achieved by selecting proper packaging materials and methods. Both effects of nonthermal processing on the packaging materials and effects of packaging materials on the quality of nonthermally processed foods during storage need to be studied for the selection of proper packaging materials and methods.

As consumers demand more nonthermally processed foods, the more id the opportunity to combine active packaging with nonthermal processing. It is expected that the use of active packaging such as MAP and antimicrobial or antioxidant films and coatings would retard the degradation of the quality of nonthermally processed food products. Thus, it is suggested to study the combination of active packaging and nonthermal processing.

References

Ahn H. J., C. Jo, J. W. Lee, J.-H. Kim, K. H. Kim & M. W. Byun. 2003. Irradiation and modified atmosphere packaging effects on residual nitrite, ascorbic acid, nitrosomyoglobin, and color in sausage. *Journal of Agricultural and Food Chemistry.* 51: 1249–1253.

Ayhan Z., H. W. Yeom, Q. H. Zhang & D. B. Min. 2001. Flavor, color, and vitamin C retention of pulsed electric field processed orange juice in different packaging materials. *Journal of Agricultural and Food Chemistry.* 49: 669–674.

Barbosa-Canovas G. V., M. M. Gongora-Nieto, U. R. Pothakamury & B. G. Swanson 1999. *Preservation of foods with pulsed electric fields.* ed. San Diego, CA: Academic Press. 197 p.

Barbosa-Canovas G. V., U. R. Pothakamury, E. Palou & B. G. Swanson 1998. *Nonthermal preservation of foods.* ed. New York: Marcel Dekker, Inc. 276 p.

Benz R. & U. Zimmermann. 1980. Pulse-length dependence of the electrical breakdown in lipid bilayer membranes. *Biochimica et Biophysica Acta.* 597: 637–642.

Buchalla R., C. Schuttler & K. W. Bogl. 1993. Effects of ionizing radiation on plastic food packaging materials: a review. Part 1. Chemical and physical changes. *Journal of Food Protection.* 56: 991–997.

Caner C., R. J. Hernandez & B. R. Harte. 2004. High-pressure processing effects on the mechanical, barrier and mass transfer properties of food packaging flexible structures: a critical review. *Packaging Technology & Science.* 17: 23–29.

Caner C., R. J. Hernandez & M. A. Pascall. 2000. Effect of high-pressure processing on the permeance of selected high-barrier laminated films. *Packaging Technology and Science.* 13: 183–195.

Caner C., R. J. Hernandez, M. Pascall, V. M. Balasubramaniam & B. R. Harte. 2004. The effect of high-pressure food processing on the sorption behaviour of selected packaging materials. *Packaging Technology & Science.* 17: 139–153.

Crook L. R. & T. D. Boylston. 2004. Flavor characteristics of irradiated apple cider during storage: Effect of packaging materials and sorbate addition. *Journal of Food Science.* 69: C557–C563.

Devlieghere F., L. Vermeiren & J. Debevere. 2004. New preservation technologies: Possibilities and limitations. *International Dairy Journal.* 14: 273–285.

Dobias J., M. Voldrich, M. Marek & K. Chudackova. 2004. Changes of properties of polymer packaging films during high pressure treatment. *Journal of Food Engineering.* 61: 545–549.

El Makhzoumi Z. 1994. "Effect of irradiation of polymeric packaging material on the formation of volatile compounds" In *Food packaging and preservation*, edited by M. Mathlouthis, pp 88–99 London: Blackie Academic & Professional.

Evrendilek G. A. & Q. H. Zhang. 2003. Effects of pH, temperature and pre-PEF treatments on PEF and heat inactivation of *Escherichia coli* O157: H7. *Journal of Food Protection.* 66: 755–759.

Goetz J. & H. Weisser. 2002. Permeation of aroma compounds through plastic films under high pressure: in-situ measuring method. *Innovative Food Science & Emerging Technologies.* 3: 25–31.

Goulas A. E., K. A. Riganakos, A. Badeka & M. G. Kontominas. 2002. Effect of ionizing radiation on the physicochemical and mechanical properties of commercial monolayer flexible plastics packaging materials. *Food Additives and Contaminants.* 19: 1190–1199.

Grahl T. & H. Markl. 1996. Killing of microorganisms by pulsed electric fields. *Applied Microbiology and Biotechnology.* 45: 148–157.

Grecz N., D. B. Rowley & A. Matsuyama 1983. "The action of radiation on bacteria and viruses" In *Preservation of foods by ionizing radiation*, edited by E. S. Josephson & M. S. Petersons, pp 167–218 Boca Raton, FL: CRC Press.

Han J., C. L. Gomes-Feitosa, E. Castell-Perez, R. G. Moreira & P. F. Silva. 2004. Quality of packaged romaine lettuce hearts exposed to low-dose electron beam irradiation. *Lebensmittel Wissenschaft und Technologie.* 37: 705–715.

Hoover D. H. 1997. Minimally processed fruits and vegetables: Reducing microbial load by nonthermal physical treatments. *Food Technology.* 51: 66–71.

Hotchkiss J. H. & M. J. Banco. 1992. Influence of new packaging technologies on the growth of microorganisms in produce. *Journal of Food Protection.* 55: 815–820.

Hulsheger H. & E. G. Niemann. 1980. Lethal effects of high voltage pulses on *E. coli* K12. *Radiation and Environmental Biophysics.* 18: 281–288.

Jin, Z. T. & Q. H. Zhang. 2002. "Cost evaluation of a commercial scale PEF system." IFT Annual Meeting, June 15–19, 2002, Institute of Food Technologists, Anaheim, California, p. 229/91E-21.

Kilcast D. 1990. "Irradiation of packaged food" In *Food irradiation and the chemist*, edited by D. E. Johnson & M. H. Stevensons, pp 140–152 U.K.: The Royal Society of Chemistry.

Kinosita K. & T. Y. Tsong. 1977. Formation and resealing of pores of controlled sizes in human erythrocyte membrane. *Nature.* 268: 438–440.

Komolprasert V., T. P. McNeal, A. Agrawal, C. Adhikari & D. W. Thayer. 2001. Volatile and non-volatile compounds in irradiated semi-rigid crystalline poly(ethylene terephthalate) polymers. *Food Additives and Contaminants.* 18: 89–101.

Krishnamurthy K., A. Demirci, V. M. Puri & C. N. Cutter. 2004. Effect of packaging materials on inactivation of pathogenic microorganisms on meat during irradiation. *Transactions of the ASAE.* 47: 1141–1149.

Kubel J., H. Ludwig, H. Marx & B. Tauscher. 1996. Diffusion of aroma compounds into packaging films under high pressure. *Packaging Technology and Science.* 9: 143–152.

Kusmider E. A., J. G. Sebranek, S. M. Lonergan & M. S. Honeyman. 2002. Effects of carbon monoxide packaging on color and lipid stability of irradiated ground beef. *Journal of Food Science*. 67: 3463–3468.

Lambert Y., G. Demazeau, A. Largeteau & J. M. Bouvier. 2000. Packaging for high-pressure treatments in the food industry. *Packaging Technology and Science*. 13: 63–71.

Lee M., J. G. Sebranek, D. G. Olson & J. S. Dickson. 1996. Irradiation and packaging of fresh meat and poultry. *Journal of Food Protection*. 59: 62–72.

Lee J. H., T. H. Sung, K. T. Lee & M. R. Kim. 2004. Effect of gamma-irradiation on color, pungency, and volatiles of Korean red pepper powder. *Journal of Food Science*. 69: C585–C592.

Marque D., A. Feigenbaum & A. M. Riquet. 1995. Consequences of polypropylene film ionisation on the food/packaging interactions. *Journal of Polymer Engineering*. 15: 101–115.

Marth E. H. 1998. Extended shelf life refrigerated foods: Microbiological quality and safety. *Food Technology*. 52: 57–62.

Masuda M., Y. Saito, T. Iwanami & Y. Hirai 1992. "Effects of hydrostatic pressure on packaging materials for food" In *High pressure and biotechnology*, edited by C. Balny, R. Hayashi, K. Heremans & P. Massons, pp 545–547 London: John Libbey Eurotext.

Matsui T., M. Inoue, M. Shimoda & Y. Osajama. 1991. Sorption of volatile compounds into electron beam irradiated EVA film in the vapour phase. *Journal of the Science of Food and Agriculture* 54: 127–135.

Min S., Z. T. Jin & Q. H. Zhang. 2003. Commercial scale pulsed electric field processing of tomato juice. *Journal of Agricultural and Food Chemistry*. 51: 3338–3344.

Min S., L. Reina & Q. H. Zhang. 2002. Effects of water activity on the inactivation of *Enterobacter cloacae* inoculated in chocolate liquor and a model system. *Journal of Food Processing and Preservation*. 26: 323–337.

Min S. & Q. H. Zhang 2005. "Packaging for non-thermal processing" In *Innovations in food packaging*, edited by J. H. Hans, pp 170–185 London: Academic Press.

Murano P. S., E. A. Murano & D. G. Olson. 1998. Irradiated ground beef: sensory and quality changes during storage under various packaging conditions. *Journal of Food Science*. 63: 548–551.

Olson D. G. 1998. Irradiation of Food. *Food Technology*. 52: 56–62.

Ozen B. F. & J. D. Floros. 2001. Effects of emerging food processing techniques on the packaging materials. *Trends in Food Science & Technology*. 12: 60–67.

Pentimalli M., D. Capitani, A. Ferrando, D. Ferri, P. Ragni & A. L. Segre. 2000. Gamma irradiation of food packaging materials: an NMR study. *Polymer*. 41: 2871–2881.

Qin B. L., U. R. Pothakamury, H. Vega, O. Martin, G. V. Barbosa-Canovas & B. G. Swanson. 1995. Food pasteurization using high intensity pulsed electric fields. *Food Technology*. 49: 55–60.

Raso J., M. L. Calderon, M. Gongora, G. V. Barbosa-Canovas & B. G. Swanson. 1998. Inactivation of mold ascospores and conidiospores suspended in fruit juices by pulsed electric fields. *Lebensmittel Wissenschaft und Technologie*. 31: 668–672.

Rojas De Gante C. & B. Pascat. 1990. Effects of β-ionizing radiation on the properties of flexible packaging materials. *Packaging Technology and Science* 3: 97–105.

Schauwecker A., V. M. Balasubramaniam, G. Sadler, M. A. Pascall & C. Adhikari. 2002. Influence of high-pressure processing on selected polymeric materials and on the migration of a pressure-transmitting fluid. *Packaging Technology and Science.* 15: 255–262.

Smith J. S. & S. Pillai. 2004. Irradiation and food safety. *Food Technology.* 58: 48–55.

Song I. H., W. J. Kim, C. Jo, H. J. Ahn, J. H. Kim & M. W. Byun. 2003. Effect of modified atmosphere packaging and irradiation in combination on content of nitrosamines in cooked pork sausage. *Journal of Food Protection.* 66: 1090–1094.

Thakur B. R., R. K. Singh & P. E. Nelson. 1996. Quality attributes of processed tomato products: A review. *Food Reviews International* 12: 375–401.

Vega-Mercado H., U. R. Pothakamury, F. J. Chang, G. V. Barbosa-Canovas & B. G. Swanson. 1996. Inactivation of *Escherichia coli* by combining pH, ionic strength and pulsed electric fields hurdles. *Food Research International.* 29: 117–121.

Wouters P. C., I. Alvarez & J. Raso. 2001. Critical factors determining inactivation kinetics by pulsed electric field food processing. *Trends in Food Science & Technology.* 12: 112–121.

Wouters P. C., A. P. Bos & J. Ueckert. 2001. Membrane permeabilization in relation to inactivation kinetics of *Lactobacillus* species due to pulsed electric fields. *Applied and Environmental Microbiology* 67: 3092–3101.

Yeom H. W., K. T. McCann, C. B. Streaker & Q. H. Zhang. 2002. Pulsed electric field processing of high acid liquid foods: a review. *Advances in Food and Nutrition Research.* 44: 1–32.

Yuste J., M. Capellas, R. Pla, D. Y. C. Fung & M. Mor-Mur. 2002. High pressure processing for food safety and preservation: A review. *Journal of Rapid Methods and Automation in Microbiology.* 9: 1–10.

Zhang Q., G. V. Barbosa-Canovas & B. G. Swanson. 1995. Engineering aspects of pulsed electric field pasteurization. *Journal of Food Engineering.* 25: 261–281.

Zhang Q., A. Monsalve-Gonzalez, B. L. Qin, G. V. Barbosa-Canovas & B. G. Swanson. 1994. Inactivation of *Saccharomyces cerevisiae* in apple juice by square-wave and exponential decay pulsed electric fields. *Journal of Food Processing and Engineering* 17: 469–478.

Zhang Q., B. L. Qin, G. V. Barbosa-Canovas & B. G. Swanson. 1995. Inactivation of *E. coli* for food pasteurization by high-strength pulsed electric fields. *Journal of Food Processing and Preservation.* 19: 103–118.

Chapter 6

PACKAGING FOR FOODS TREATED BY IONIZING RADIATION

Vanee Komolprasert

Disclaimer: The information and conclusions presented in this book chapter do not represent new agency policy nor do they imply an imminent change in existing policy.

Introduction

Ionizing radiation is a nonthermal process utilized to achieve the preservation of food. At a maximum commercial irradiation dose of 10 kGy, irradiation does not impart heat to the food, and the nutritional quality of the food is generally unaffected. The irradiation process can reduce microbial contamination of food, resulting in improved microbial safety as well as in extended shelf life of the food. In the last decade, many studies have been conducted on irradiation of various foods, especially foods susceptible to food-borne outbreaks, such as meat and meat products. Nonetheless, from a commercial standpoint, foods are generally prepackaged in the final form (a.k.a. case ready) before irradiation to avoid recontamination.

Title 21 of the Code of Federal Regulations (CFR), Part 179.25 (general provisions for food irradiation), subparagraph (c), states that packaging materials subjected to irradiation incidental to the radiation treatment and processing of prepackaged foods shall comply with 21 CFR 179.45 (packaging materials for use during the irradiation of prepackaged foods).

This list in 21 CFR 179.45 does not include many modern packaging materials presently desired by the food industry in light of the commercialization of food irradiation. This, in turn, has presented new challenges to the FDA and regulated industry. This book chapter briefly describes: (i) the food additive regulations pertaining to food and packaging materials in contact with food during irradiation, (ii) emerging research aimed at determining the radiolysis products (RPs) formed from new packaging materials, including polymers and additives, after exposure to irradiation, and (iii) approaches to evaluate the pre-market safety assessment of new packaging materials in contact with food during irradiation.

US Regulations for Irradiation of Food and Packaging in Contact with Food

Under Section 201(s) of the Food Additives Amendment to the Federal Food, Drug, and Cosmetic Act (the Act) of 1958, the definition of a food additive includes ". . . any source of radiation intended for such use." In connection with Section 409 of the Act, this means that a food is deemed adulterated and cannot be legally marketed if it has been intentionally irradiated, unless the irradiation is carried out in compliance with an applicable regulation under the prescribed conditions of use specified in the regulation. The use of packaging materials for irradiated food is considered a new use and is subject to pre-market safety evaluation. Components of packaging materials that have been irradiated may migrate to food at different levels in comparison to unirradiated materials. This also holds true for the RPs produced as a result of irradiation of the packaging materials. This safety concern can be considered in two scenarios—either a packaging material is irradiated before food contact or it is irradiated while in direct contact with food (FDA 1986). In either case, the packaging material used should comply with the applicable regulations, as specified in Part 179.45. Thus, as dictated by Chapter 21 of the CFR, Part 179.25 (denoted as 21 CFR 179.25), the irradiation of both food and packaging materials in contact with food is subject to pre-market approval before introduction of the food into interstate commerce.

21 CFR 179 is the primary regulation that covers irradiation in the production, processing, and handling of food, and it is divided into subparts and sections as shown in Table 6.1. Subpart B describes radiation and radiation sources, which include gamma ray, electron beam (e-beam), and X-ray, as well as the general provisions for food

Table 6.1.   Sections under 21 CFR Part 179—irradiation in the production, processing, and handling of food.

179.21	Sources of radiation used for inspection of food, for inspection of packaged food, and for controlling food processing
179.25	General provisions for food irradiation
179.26	Ionizing radiation for the treatment of food
179.30	Radiofrequency radiation for the heating of food, including microwave frequencies
179.39	Ultraviolet radiation for the processing and treatment of food
179.41	Pulsed light for the treatment of food
179.45	Packaging materials for use during the irradiation of prepackaged foods

irradiation. Subpart B also lists other radiation processes, including radiofrequency radiation, ultraviolet, and pulsed light. These radiation processes are covered elsewhere and will not be included in this chapter. Subpart C describes packaging materials for irradiated foods. The listing regulations in Part 179 are the result of approvals through the food additive petition process codified in 21 CFR 171.

Subpart B includes 21 CFR 179.26(b) that lists foods currently permitted to be irradiated (shown in Table 6.2). Irradiated food should be adequately labeled under the general labeling requirements in 21 CFR 179.26(c).

General Aspects of Irradiation and Food Irradiation

Ionizing radiation for the treatment of packaged food can be achieved using gamma rays (with cobalt-60 or cesium-137 radioisotope), electron beams, or X-rays, as specified in 21 CFR 179.26(a). The effects of radiation on matter generally depend on the type of the radiation and energy level, as well as on the composition, physical state, temperature, and environment of the absorbing material, whether it is food or the packaging materials in contact with the food. Chemical changes in matter can occur via primary radiolysis effects, which occur as a result of the adsorption of the energy by the absorbing matter, and can have biological consequences in the case where the target materials include living organisms. With proper application, irradiation can be an effective means of eliminating and/or reducing the microbial load and thus the food-borne diseases they induce, thereby improving the safety of many foods as well as extending their shelf life.

Table 6.2. Foods Permitted to be Irradiated Under FDA's Regulations (21 CFR 179.26)

Food	Purpose	Dose
Fresh, nonheated processed pork	Control of *Trichinella spiralis*	0.3 kGy minimum to 1 kGy maximum
Fresh foods	Growth and maturation inhibition	1 kGy maximum
Foods	Arthropod disinfection	1 kGy maximum
Dry or dehydrated enzyme preparations	Microbial disinfection	10 kGy maximum
Dry or dehydrated spices/seasonings	Microbial disinfection	30 kGy maximum
Fresh or frozen, uncooked poultry products	Pathogen control	3 kGy maximum
Frozen packaged meats (solely NASA)	Sterilization	44 kGy minimum
Refrigerated, uncooked meat products	Pathogen control	4.5 kGy maximum
Frozen uncooked meat products	Pathogen control	7 kGy maximum
Fresh shell eggs	Control of *Salmonella*	3.0 kGy maximum
Seeds for sprouting	Control of microbial pathogens	8.0 kGy maximum
Fresh or frozen molluscan shellfish[a]	Control of *Vibrio* species and other food-borne pathogens	5.5 kGy maximum

[a]FDA 2005.

Expert groups of national and international organizations as well as many regulatory agencies have generally concluded that irradiated food is safe and wholesome and that food irradiation at commonly used dosing levels does not present any enhanced toxicological, microbiological, or nutritional hazards to the food beyond those brought about by conventional food-processing techniques. These experts have agreed that irradiation of food for microbial safety should be carried out under good manufacturing practices (GMPs) and good irradiation practices (GIPs). Subsequently, standards on various aspects of radiation processing have been developed and internationally accepted (Farrar *et al.* 1993).

The World Health Organization (WHO) considers ionizing radiation an important process toward ensuring food safety (Diehl 1995). It can be a useful control measure in the production of several types of raw or minimally processed foods such as poultry, meat and meat products, fish, seafood, and fruits and vegetables (Molins, Motarjemi, and –Käferstein 2001). An increased interest in food irradiation for quality and microbiological safety was realized by several emerging studies on various food products, including irradiation of meat and meat products (Lacroix *et al.* 2000, 2002, 2004; Montgomery *et al.* 2003; Ahn and Lee 2004; Ahn and Nam 2004), processed meat and ready-to-eat foods (Jo *et al.* 2003; Sommers *et al.* 2003, 2004a, 2004b; Chawla and Chander 2004; Cava *et al.* 2005), poultry products (Nam and Ahn 2003; Lacroix and Chiasson 2004; Javanmard *et al.* 2006), chicken eggs (Pinto *et al.* 2004), seafood (Andrews and Grodner 2004), fresh produce (Chaudry *et al.* 2004; Goularte *et al.* 2004; Han *et al.* 2004; Martins *et al.* 2004; Lacroix and Lafortune 2004; Prakash and Foley 2004), sprouts (Rajkowski and Fan 2004), and fruit juices (Fan *et al.* 2004).

Packaging Materials for Prepackaged Irradiated Foods

Present-day food processors prefer that food be prepackaged in the final packaging form before irradiation to prevent recontamination and to facilitate prompt shipment to market after irradiation. Food could potentially become contaminated with RPs formed in the packaging materials when irradiated in contact with food. This may lead to a safety concern, and, therefore, testing of packaging materials after exposure to irradiation is an integral part of the pre-market safety assessment of packaging materials irradiated in contact with food.

Irradiation can cause changes to the packaging material that might affect its integrity and functionality as a barrier, for example, to chemical or microbial contamination. Most food packaging materials are composed of polymers that may be susceptible to chemical changes induced by ionizing radiation which are the result of two competing reactions, cross-linking (polymerization) and chain scission (degradation). Radiation- induced cross-linking of polymers dominates under vacuum or an inert atmosphere. Chain scission dominates during irradiation of polymers in the presence of oxygen or air. Both reactions are random, are generally proportional to the dose, and depend on the dose rate and the oxygen content of the atmosphere in which the polymer is irradiated. The idea of cross-linking predominating under vacuum or an inert atmosphere is important because it has served as the basis for recent approvals under 21 CFR 170.39 (see below) for packaging materials irradiated in contact with food under nonoxygen atmospheres (FDA 2006a).

The RPs formed upon irradiation of a polymer or adjuvant could migrate into food and affect odor, taste, and safety of the irradiated food (Deschênes *et al.* 1995; Welle, Mauer, and Franz 2002; Franz and Welle 2004; Stoffers *et al.* 2004). Radiation does not generally affect all properties of a polymer to the same degree. Therefore, the effect of radiation on the formation of RPs and the degree to which the RPs (as well as the base packaging materials) become components of an individual's daily diet under the intended conditions of use must be determined.

Packaging materials irradiated in contact with food are subject to pre-market approval by the FDA and may be used only if they comply with the regulation listings in 21 CFR 179.45 (as discussed above) and are the subject of an effective food contact notification (FCN) or a threshold of regulation (TOR) exemption under 21 CFR 170.39 (as described in 21 CFR 179.25(c)). These three regulatory options available to all food contact substances (FCSs) are discussed in detail by Twaroski *et al.* (2006). Table 6.3 lists the packaging materials presently authorized under 21 CFR 179.45 and includes films and homogeneous structures at various doses, most of which were initially approved several years ago for irradiation by gamma-ray treatment. In response to a recent food additive petition, the listed materials under 21 CFR 179.45(b) were evaluated for being subjected to a dose not to exceed 10 kGy incidental to the use of any radiation source in the treatment of packaged food. In its safety assessment of that petition, FDA concluded that gamma-ray, e-beam, and X-ray sources are equivalent in terms of the types and levels of RPs generated in the packaging materials under the conditions at which prepackaged foods are irradiated (FDA 2001). As a result, 21 CFR 179.45(b) was recently amended to allow the listed materials to be subjected to a dose not to exceed 10 kGy incidental to the use of any radiation source in the radiation treatment of prepackaged foods. Recently, submissions for specific packaging constructions and conditions of use were exempted from the need for a regulation listing as per 21 CFR 170.39 (FDA 2006a).

It should be noted that the packaging materials in 21 CFR 179.45 do not generally meet today's needs as do newer materials that may be

Table 6.3.   Packaging materials listed in 21 CFR 179.45 for use during irradiation of prepackaged foods.

Packaging Materials	Maximum Dose (kGy)
Section 179.45(b)	
Nitrocellulose-coated cellophane	10
Glassine paper	10
Wax-coated paperboard	10
Polyolefin film	10
Kraft paper	0.5
Polyethylene terephthalate film (basic polymer)	10
Polystyrene film	10
Rubber hydrochloride film	10
Vinylidene chloride–vinyl chloride copolymer film	10
Nylon 11 [polyamide-11]	10
Section 179.45(c)	
Ethylene–vinyl acetate copolymer	30
Section 179.45(d)	
Vegetable parchment	60
Polyethylene film (basic polymer)	60
Polyethylene terephthalate film	60
Nylon 6 [polyamide-6]	60
Vinyl chloride–vinyl acetate copolymer film	60

more desirable to the food industry. However, many of these newer packaging materials have not yet been evaluated by FDA. In addition to the base polymers, adjuvants such as anitoxidants and stabilizers are also of concern with regard to RPs. Such adjuvants are prone to degradation during polymer processing and, moreover, during irradiation as they degrade preferentially over the polymer and could result in significant levels of RPs migrating into food. Therefore, the migration of both base polymers and adjuvants, as well as migration of their RPs, must be evaluated in the pre-market safety assessment prior to their use.

Evaluation of New Packaging Materials for Irradiation in Contact with Food

To demonstrate the radiation stability of a packaging material, it must be shown that irradiation does not significantly alter the physical and chemical properties of the material (Chuaqui-Offermanns 1989). Physical and chemical testing of materials is widely acceptable for radiation sterilization of medical products, in which their properties are important for the desired performance of the medical products during use. This type of testing is not sufficient for packaging materials intended to contact food during irradiation. Packaging materials and their RPs can potentially migrate into food at significant levels, although changes in the material properties of the packaging may be insignificant or undetected. Hence, the overall assessment of new packaging materials in contact with food during irradiation should include an assessment of any changes in physical and chemical properties as a result of irradiation, and a pre-market safety assessment of the packaging materials and their RPs that might become components of an individual's diet under the conditions of use.

Most of the early studies that described the effects of ionizing radiation on various food packaging polymers, such as polyolefins, were conducted at fairly high dose levels compared to the levels typically used in food processing. Most of the new studies have extended irradiation to other homopolymers, as well as to copolymer and multilayer structures and adjuvants. Paquette (2004) surveyed the literature for the types of RPs formed and their respective concentrations in several irradiated polymers, including PS, PET, LDPE, PP, EVA, PA6, and PVC (abbreviations are listed in the Appendix), all of which contained adjuvants, as well as concentrations of the RPs in food simulants. Paquette concluded that the formation of RPs depended on the absorbed dose, dose rate, atmosphere, temperature, time after irradiation, and food simulant. The RPs from the polymers consisted of low-molecular-weight aldehydes, acids, and olefins. The dietary exposures to most RPs formed in the materials surveyed were determined to be less than $0.5\,\mu g/kg$ in the daily diet, less than the TOR concern level as per 21 CFR 170.39. However, the author noted that in contrast to the base polymers, the adjuvants identified in the survey were not currently listed in 21 CFR 179.45.

Polyolefins, EVOH, and Polyamides

Loveridge and Milch (2004) recently demonstrated that physical testing is necessary for determining the integrity of multilayered food pouches irradiated at a very high dose >44 kGy). The authors observed that the irradiation significantly contributes to seal strength loss that affects the package integrity. Goulas *et al.* (2004a) observed that gamma irradiation >30-kGy dose discolored most monolayer and multilayer commercial semirigid materials made of PS, PP, PET, PVC/HDPE, HDPE/PA, and HDPE. Goulas and co-workers (2003) reported that lower dose levels (5- to 10-kGy gamma radiation) did not significantly change the mechanical and permeation properties as well as overall migration levels of commercial multilayer films, including co-extruded PP, EVOH, LDPE, LLDPE, PA, and ionomer. Chytiri *et al.* (2006) experimentally showed that the effects of 5- to 60-kGy gamma radiation on the thermal, mechanical, and permeation properties, as well as infrared spectra, of multilayered films containing 25–50 wt% of a recycled LDPE layer sandwiched between virgin LDPE layers, were not different from those of films made with 100% virgin LDPE. Regardless of the presence of recycled LDPE in the film, the authors concluded that 60-kGy irradiation dose induced mechanical change but did not affect other properties of the film.

Deschênes *et al.* (1995) observed that a PA/PVDC/EVA multilayered barrier film produced volatile aldehydes and other hydrocarbons that caused off-odor and taints in water irradiated as low as 1 kGy. Riganakos *et al.* (1999) detected volatile compounds generated in monolayer films of LDPE and EVA copolymer and multilayer films (PET/PE/EVOH/PE) after e-beam radiation at 5, 20, and 100 kGy doses. These volatiles were primary and secondary oxidation products including aldehydes, ketones, alcohols, and carboxylic acids, all of which are known to affect the organoleptic properties and shelf life of irradiated foods. The concentrations of these compounds increased with increased irradiation dose. Regardless of the irradiation dose, the authors concluded that the infrared spectra as well as the permeability to oxygen, water, and carbon dioxide of these materials were not significantly altered by irradiation even at doses as high as 100 kGy.

Adjuvants

Several recent studies focused on the effects of irradiation on hindered phenol antioxidants in packaging polymers. Deschênes *et al.* (2004) reported that gamma radiation degraded a hindered phosphite antioxidant, Irgafos 168, in PE and PP to phosphate oxidation products. The phosphate products further degraded to other products during irradiation. Similarly, Stoffers and co-workers (2004) observed that irradiation degraded Irganox 1076 and Irgafos 168 in PE but did not affect Irganox 1076 in PS even after irradiation at a dose up to 54 kGy. This was reported to be indicative of the stability of PS, which was also observed by Kawamura (2004). The authors also reported that adjuvant RPs and other degradation products affected the sensory properties of the irradiated LDPE, HDPE, PA6, and PA12.

Jeon *et al.* (2007) did not detect migration of Irgafos 168 from a LLDPE pouch irradiated up to 200 kGy in food simulants (distilled water, 4% (v/v) acetic acid, and 20% aqueous ethanol). However, the authors detected decreasing migration levels of Irganox 1076 with increased irradiation dose. On the other hand, levels of the Irganox 1076 RPs (2,4-di-tert-butylphenol, 1, 3-di-tert-butylbenzene, and toluene) in the simulants increased with increased irradiation dose. These common RPs were also detected by Kawamura (2004) who extensively studied the effects of 10- to 50-kGy gamma radiation on numerous antioxidants and UV stabilizers in commercial sheets and films of PE, PP, and PS. The author concluded that polymer stability was in the order of PS > PE > PP. These results are consistent with a study of Pentimalli and co-workers (2000), which showed that PS was not affected by gamma irradiation even at high doses. Kawamura also observed that the presence of adjuvants reduced the loss of mechanical properties of the polymers and were essential during irradiation of the polymers. The authors also studied the effects of irradiation on polybutylene and some copolymers in the absence or presence of antioxidant and stabilizers and observed the crucial role of antioxidants and stabilizers in protecting polybutylene and butadiene-containing copolymers from degradation reactions.

Plasticizers

Goulas *et al.* (2004b) extensively studied the effects of both gamma and e-beam radiation, at doses in a range of 4–50 kGy, on migration of the plasticizers di-(2-ethylhexyl)adipate (DEHA) and acetyltributyl citrate (ATBC) from PVC and PVDC/VC copolymer films into chicken meat and olive oil. They found that 4- to 9-kGy radiation did not affect migration of these plasticizers into food but that dose levels of 20- to 50-kGy radiation did lead to higher levels in food. Zygoura *et al.* (2007) studied the effects of gamma radiation, in the range of 10–25 kGy, on the migration of DEHA and ATBC from a PVC film into isooctane, a solvent often used as a fatty food simulant. The authors reported that plasticizer migration levels increased with increased gamma radiation. Unlike Irgafos 168, irradiation did not induce transformations of these two plasticizers.

Semirigid PET

PET films are regulated in 21 CFR 179.45, but semirigid and rigid PET structures are not. The effects of irradiation on semirigid and rigid PET structures may be different from those on the films. Komolprasert (1998) conducted a preliminary study using ten different semirigid PET materials (crystalline and amorphous) of various properties and characteristics, irradiated at 25-kGy gamma radiation, which are listed in Table 6.4. The nonvolatile, soluble solids were extracted from these PET materials, using Soxhlet extraction with acetone, and the percent extractable results are presented in Table 6.5. These results indicated that 25-kGy gamma irradiation did not significantly increase the amount of soluble solids or PET cyclic trimer. PET cyclic trimer was the major component that accounted for more than 50% of the total soluble solids extracted.

Komolprasert and co-workers (2001) conducted an in-depth study on the effects of a 25-kGy gamma radiation on two of the semirigid crystalline PET polymers listed in Table 6.4 (categories 1 and 5). The authors reported that the volatiles detected were formic acid, acetic acid, 1,3-dioxalane, and 2-methyl-1,3-dioxolane. As discussed above, PET cyclic trimer was the major nonvolatile detected in the soluble solids, and the level was not affected by irradiation. The authors concluded that irradiation of PET at 25 kGy significantly increased the amount of volatiles but not the nonvolatiles.

Komolprasert and co-workers (2003a) continued their study using two of the semirigid amorphous PET copolymers listed in Table 6.4 (categories 9 and 10) irradiated using both gamma and e-beam radiation at 5, 25, and 50 kGy. The authors reported detection of the same volatiles as in the previous study, but acetaldehyde, 1,3-dioxolane, and 2-methyl-1,3-dioxolane were quantifiable. Levels of these volatiles increased with increased radiation dose. Besides PET cyclic trimer, terephthalic acid, bis-(2-hydroxyethyl) terephthalate, dimethyl terephthalate, monohydroxy ethylene terephthalate, and tetramer are additional nonvolatiles that are generally known to be present in the soluble fraction from PET. Levels of these nonvolatiles did not significantly

Table 6.4. Density and percent crystallinity of various semirigid PET materials.

Material Type	Cat. No.	A vs. C	O vs. N	Density	% X	Thickness (mm)	Material Characteristics
PET homopolymer	1	C	O	1.36	25	0.3	Sheet, IV 0.72
	2	A	N	1.34	2	0.6	Sheet, IV 0.95
	3	C	N	1.34	N/A	0.7	Tray,1.5 mole % DEG, 3 wt% polyolefin, IV 0.95
	4	C	N	1.38	36	0.6	Sheet, IV 0.95
IPA copolymer	5	C	O	1.37	30	0.3	Sheet, 1.5 mole % IPA, IV 0.8
	6	A	N	1.34	0	0.5	Sheet, 4 mole % IPA, IV 0.8
	7	C	N	1.37	39	0.5	Sheet, 4 mole % IPA, IV 0.8
CHDM copolymer	8	C	O	1.37	29	0.4	Bottle, 1.5 mole % CHDM, IV 0.8
	9	A	N	1.33	5	0.3	Sheet, 3.0 mole % CHDM, 1.5 wt% DEG, IV 0.8
PETG copolymer	10	A	N	1.33	Low	0.3	Sheet, 31 mole % CHDM, IV 0.73

Cat. No. 1, 2, 4, 5–7 supplied by, formerly called, Shell Chemical Company. Cat. No. 3, 8–10 supplied by Eastman Chemical Company. A, amorphous; C, crystalline; O, oriented and stretched; N, non-oriented; X, crystallinity; IPA, isophthalic acid; CHDM, 1,4-cyclohexane dimethanol; DEG, diethylene glycol; PETG, polyethylene terephthalate, glycol modified.

Table 6.5. Percents of soluble solid extracted from PET test materials before (NIR) and after 25-kGy gamma irradiation (IR).

Category Number	# Rep.	% Soluble Solid		% PET Cyclic Trimer in Soluble Solid	
		NIR	IR	NIR	IR
1	5	0.78 ± 0.07	$(0.78 \pm 0.06)^{IS}$	0.50 ± 0.01	$(0.50 \pm 0.01)^{IS}$
2	5	0.56 ± 0.04	$(0.58 \pm 0.02)^{IS}$	0.28 ± 0.02	$(0.28 \pm 0.02)^{IS}$
3	5	0.48 ± 0.06	$(0.49 \pm 0.05)^{IS}$	0.26 ± 0.05	$(0.28 \pm 0.02)^{IS}$
4	5	0.48 ± 0.02	$(0.49 \pm 0.06)^{IS}$	0.25 ± 0.06	$(0.22 \pm 0.07)^{IS}$
5	5	0.67 ± 0.05	$(0.68 \pm 0.05)^{IS}$	0.41 ± 0.01	$(0.41 \pm 0.02)^{IS}$
6	5	0.76 ± 0.05	$(0.78 \pm 0.06)^{IS}$	0.38 ± 0.06	$(0.38 \pm 0.06)^{IS}$
7	5	0.48 ± 0.08	$(0.49 \pm 0.06)^{IS}$	0.23 ± 0.01	$(0.21 \pm 0.01)^{IS}$
8	5	0.64 ± 0.07	$(0.66 \pm 0.03)^{IS}$	0.38 ± 0.09	$(0.36 \pm 0.02)^{IS}$
9	5	0.77 ± 0.05	$(0.80 \pm 0.07)^{IS}$	0.37 ± 0.06	$(0.38 \pm 0.04)^{IS}$
10	5	0.93 ± 0.02	$(0.92 \pm 0.05)^{IS}$	0.51 ± 0.01	$(0.51 \pm 0.02)^{IS}$

IS, insignificant at $P < 0.05$.

increase after irradiation at all doses. The authors also conducted migration studies on these polymers using 10% aqueous ethanol and 100% heptane solvents under various conditions of use (i.e., FDA migration testing protocols). Migration levels of these nonvolatiles were less than 5 ppb under conditions of Use E (room temperature filled and stored) and F (refrigeration storage), with levels higher than 5 ppb under condition of Use D (hot filled or pasteurized below 66°C). Based on the overall results obtained, the authors concluded that gamma and e-beam radiation did not generate any new, nonvolatile RPs, and radiation effects on these PET materials are similar, regardless of the irradiation dose.

Nylon 6I/6T

Nylon 6I/6T, a polyamide of 1,6-hexamethylene, terephthalic acid, and isophthalic acid, is also among the new packaging materials of interest to industry for use in contact with food during irradiation. Nylon 6I/6T of certain compositions is listed for certain uses in 21 CFR 177.1500(a)(12), but it is not authorized for contact with food during irradiation. Komolprasert et al. (2002) and McNeal et al. (2004) studied the effects of gamma irradiation on Nylon 6I/6T powder (amorphous, 1.2 g/cc) at doses up to 50 kGy. The authors reported that irradiation generated volatiles, that is, n-butanal, acetic acid, methyl-cyclopentene-1-one, methyl ethyl ketone, and *n*-pentanal, at levels as high as 2, 30, 2, 2, and 72 ppm, respectively, in the irradiated powder samples. The authors evaluated nonvolatiles by determining the percent soluble solids extracted from the powder specimens before and after 5- to 50-kGy irradiation, using 10 and 50% ethanol solutions at 40°C for 10 days. As shown in Table 6.6, irradiation did not significantly increase the percent soluble solids from the powders.

Figure 6.1 contains the representative HPLC chromatograms of the 10% ethanol extractables for 50-kGy irradiated and nonirradiated (NIR)

Table 6.6. Percent soluble solids extracted from PA powders after gamma irradiation at 0 (NIR), 5, 25, and 50 kGy (IR), using 10% ethanol and 50% ethanol in water maintained at 40°C for 10 days.

	# Rep.	NIR	5 kGy	25 kGy	50 kGy
10% ETOH	5–6	0.49 ± 0.10	$(0.58 \pm 0.10)^{IS}$	$(0.46 \pm 0.08)^{IS}$	$(0.55 \pm 0.11)^{IS}$
50% ETOH	5–6	0.97 ± 0.18	$(1.03 \pm 0.26)^{IS}$	$(0.97 \pm 0.12)^{IS}$	$(1.05 \pm 0.09)^{IS}$

IS, insignificant difference for IR vs. NIR at $P < 0.05$.

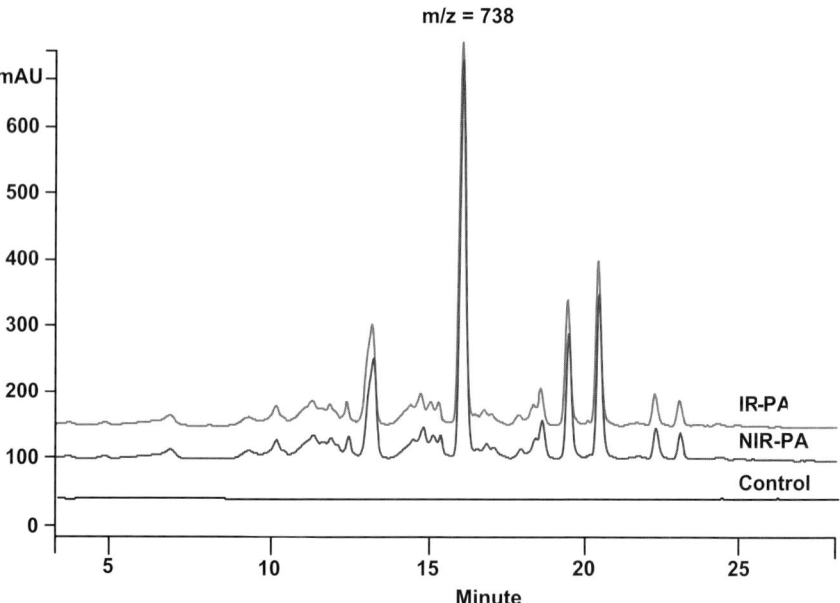

Figure 6.1. Representative liquid chromatograms of extractable solids from nonirradiated (NIR) and irradiated powdered PA test specimens (IR). Extraction conditions: 10% ethanol/water maintained at 40°C for 10 days. HPLC/PDA analysis at 210 nm. $m/z = 738$ represents the parent ion consistent with an oligomer of $n = 3$.

Nylon 6I/6T powder. The chromatograms are not significantly different, suggesting that irradiation had no effect on the formation of non-volatiles present in the ethanol extractable solids. The largest peak, with a parent ion $m/z = 738$ (verified by LC/MS), was consistent with a Nylon 6I/6T oligomer for $n=3$. The overall results indicated that irradiation significantly increased the concentrations of volatiles but had little or no measurable effect on the extractable nonvolatiles.

Ethylene-Vinyl Alcohol Copolymer (EVOH)

EVOH copolymers listed in 21 CFR 177.1360 are widely used as a nonfood contact barrier layer in food packaging. They significantly improve the gas- and vapor-barrier properties of multilaminate food packages. EVOH copolymers are not currently approved for use in contact with food during irradiation. Komolprasert *et al.* (2003b) and McNeal *et al.* (2004) studied the effects of 5- to 50-kGy e-beam irradiation on an EVOH copolymer powder (38 mol % ethylene, 1.17 g/cc) manufactured with and without α-methyl styrene dimer as an inhibitor.

For volatiles, the authors reported that e-beam radiation produced many low-molecular-weight aliphatic hydrocarbons and oxidation products in EVOH powder without inhibitor, while fewer polymer breakdown products were detected in EVOH powder with inhibitor. For EVOH powder containing inhibitor, the major volatiles were breakdown products of the inhibitor, including alkyl aromatics such as tert-butyl benzene and isobutyl benzene, and oxygenated aromatics such as acetophenone and 2-phenyl isopropanol. The results, reported in Table 6.7, indicate that e-beam irradiation also produced propanal, methyl ethyl ketone, 2-butanol, tert-butyl benzene, and 2-methylpropylbenzene in measurable amounts. Volatiles detected in both NIR and irradiated test specimens were acetic acid, cumene, α-methyl styrene, acetophenone, and 2-phenyl isopropanol. The authors concluded that the concentrations of most volatile substances in the EVOH powder samples increased with the irradiation dose, but the increase in concentration was nonlinear.

For nonvolatiles, the authors observed that the percent extractable soluble solids from EVOH powder was greater with 50% ethanol than with 10% ethanol (Tables 6.8 and 6.9). For EVOH powder with inhibitor, e-beam irradiation at 5–50 kGy doses did not significantly increase the levels of extractable solids soluble in 10 and 50% ethanol (Table 6.8). On the other hand, for EVOH powder without inhibitor, e-beam radiation significantly increased the levels of extractable soluble

Table 6.7. Concentrations (μg/g) of volatiles present in nonirradiated (NIR) EVOH (with inhibitor) powders and after 5, 25, and 50 kGy e-beam irradiation (IR).

Volatile	# Rep.	NIR	5 kGy	25 kGy	50 kGy
Propanal	4–6	<0.2	$(0.28 \pm 0.05)^S$	$(0.5 \pm 0.08)^S$	$(0.54 \pm 0.07)^S$
Methyl ethyl ketone	4–6	<0.2	$(3.0 \pm 0.23)^S$	$(2.9 \pm 0.21)^S$	$(5.1 \pm 0.79)^S$
2-Butanol	4–6	<0.1	$(4.1 \pm 0.35)^S$	$(6.1 \pm 0.61)^S$	$(7.0 \pm 0.85)^S$
Acetic acid (IR/NIR)	4–6	1	1.33	1.97	2.75
Cumene	4–6	0.72 ± 0.06	$(7.5 \pm 0.56)^S$	$(7.2 \pm 0.51)^S$	$(4.8 \pm 0.41)^S$
α-Methyl styrene	4–6	2.1 ± 0.07	$(1.3 \pm 0.1)^S$	$(0.59 \pm 0.06)^S$	$(0.45 \pm 0.06)^S$
Tert-butyl benzene	4–6	<0.1	$(4.4 \pm 0.44)^S$	$(2.3 \pm 0.21)^S$	$(1.4 \pm 0.19)^S$
2-Methyl propyl benzene	4–6	<0.1	$(1.3 \pm 0.11)^S$	$(0.64 \pm 0.07)^S$	$(0.37 \pm 0.06)^S$
Acetophenone	4–6	0.68	$(7.6 \pm 0.83)^S$	$(5.6 \pm 0.40)^S$	$(3.8 \pm 0.55)^S$
2-Phenyl-2-propanol	4–6	1.08 ± 0.07	$(4.6 \pm 0.45)^S$	$(6.1 \pm 1.1)^S$	$(4.9 \pm 0.65)^S$

S, significant at $P < 0.05$ (IR vs. NIR).
Acetic acid concentrations represented in ratios (IR/NIR); 1 = no change nor increase.

Table 6.8. Percent soluble solids extracted from nonirradiated (NIR) EVOH (with Inhibitor) powders and after e-beam irradiation at 5, 25, and 50 kGy, using aqueous 10% ethanol and 50% ethanol solution maintained at 40°C for 2, 5, and 10 days.

Food Simulant	Day	NIR	5 kGy	25 kGy	50 kGy
10% ETOH					
	2	0.43 ± 0.20	$(0.36 \pm 0.09)^{IS}$	$(0.24 \pm 0.03)^{S}$	$(0.47 \pm 0.21)^{IS}$
	5	0.45 ± 0.10	$(0.47 \pm 0.08)^{IS}$	$(0.50 \pm 0.17)^{IS}$	$(0.41 \pm 0.10)^{IS}$
	10	0.64 ± 0.09	$(0.74 \pm 0.06)^{IS}$	$(0.52 \pm 0.21)^{IS}$	$(0.62 \pm 0.03)^{IS}$
50% ETOH					
	2	1.07 ± 0.04	$(1.15 \pm 0.15)^{IS}$	$(1.79 \pm 0.54)^{IS}$	$(1.79 \pm 0.05)^{IS}$
	5	1.97 ± 0.39	$(1.90 \pm 0.75)^{IS}$	$(1.36 \pm 0.66)^{IS}$	$(1.79 \pm 0.54)^{IS}$
	10	2.75 ± 0.32	$(2.15 \pm 0.33)^{IS}$	$(2.23 \pm 0.45)^{IS}$	$(2.52 \pm 0.72)^{IS}$

IS, insignificant at $P < 0.05$ (IR vs. NIR); S, significant at $P < 0.05$ (IR vs. NIR).

Table 6.9. Percent soluble solids extracted from nonirradiated (NIR) EVOH (without inhibitor) powders and after e-beam irradiation at 5, 25, and 50 kGy, using aqueous 10% ethanol and 50% ethanol solution maintained at 40°C for 2, 5, and 10 days.

Food Simulant	Day	NIR	5 kGy	25 kGy	50 kGy
10% ETOH					
	2	0.27 ± 0.05	$(0.29 \pm 0.03)^{IS}$	$(0.33 \pm 0.02)^{IS}$	$(0.37 \pm 0.01)^{S}$
	5	0.30 ± 0.02	$(0.35 \pm 0.01)^{S}$	$(0.37 \pm 0.02)^{S}$	$(0.41 \pm 0.01)^{S}$
	10	0.30 ± 0.01	$(0.33 \pm 0.02)^{IS}$	$(0.35 \pm 0.01)^{S}$	$(0.39 \pm 0.02)^{S}$
50% ETOH					
	2	1.12 ± 0.06	$(1.25 \pm 0.16)^{S}$	$(1.54 \pm 0.06)^{S}$	$(1.56 \pm 0.15)^{S}$
	5	1.22 ± 0.11	$(1.32 \pm 0.20)^{IS}$	$(1.76 \pm 0.25)^{S}$	$(1.72 \pm 0.20)^{S}$
	10	1.24 ± 0.06	$(1.53 \pm 0.20)^{S}$	$(1.84 \pm 0.18)^{S}$	$(1.94 \pm 0.20)^{S}$

IS, insignificant at $P < 0.05$ (IR vs. NIR); S, significant at $P < 0.05$ (IR vs. NIR).

solids in both 10 and 50% ethanol (Table 6.9). The authors concluded that EVOH powder containing α-methyl styrene dimer was chemically more stable than EVOH without α-methyl styrene dimer when exposed to e-beam radiation up to 50 kGy.

An Adsorbent Pad

Another packaging material that is integral to prepackaged food is the absorbent pad. The adsorbent pad is widely used for refrigerated, uncooked meat, poultry, pork, and seafood products. Komolprasert (1999) conducted a preliminary study (unpublished) to determine the effects of 7-kGy gamma irradiation on an adsorbent pad. In the study, the absorbent pad was comprised of two white-pigmented LDPE layers, one of which was perforated, which were then sealed on all four sides to contain a cellulose pad. The identity of additives and adhesives used in the LDPE layers of the adsorbent pad was unavailable. Test pads were exposed to 7-kGy gamma irradiation at room temperature, analyzed for volatile and semivolatiles using HS/GC/MS, and subsequently extracted for nonvolatiles with 10% ethanol and 2-propanol at 40°C for 10 days.

The initial HS/GC/MS results showed that the irradiation generated 2.3–3.14 ppm of 1,3-di-tert-butyl benzene, 0.78–1.40 ppm of nonanal, and 0.24–0.41 ppm of cyclopentanone based on the adsorbent pad weight. 1,3-Di-tert-butylbenzene was a degradation product of hindered phenolic and arylphosphite antioxidants (Marque *et al.* 1998; Krzymein *et al.* 2001). Other volatiles detected at lower levels by GC/MS were acetophenone, 2-butoxyethanol, 1-cyclopentylethanone, and benzaldehyde.

The solvent extraction results showed that the soluble extractable solids using 10% ethanol increased from 0.28% before irradiation to 0.43% after irradiation, and the solids in 2-propanol increased slightly from 0.84% before irradiation to 0.91% after irradiation. Qualitative HPLC/PDA analysis of the soluble solids showed that irradiation did not generate any new substances, rather the residues consisted of BHT, Irganox 1010, Irganox 1076, and 2,4-di-tert-butylphenol (largest peak), a byproduct from the degradation of phenolic antioxidants such as BHT, Irganox 1010, and Irgafos 168 (Marque *et al.* 1998; Carlsson *et al.* 2001; Krzymein *et al.* 2001), which are commonly used in LDPE (Ehret-Henry *et al.* 1992).

Additional HPLC analyses were performed to determine degradation products in the cellulose pad. The results showed that irradiation did not generate significant amounts of glucose and cellobiose, as judged by their migration levels in 10% ethanol (<10 ppb) after a storage at 40°C for 10 days. Irradiation could produce hydrolyzed cellulose material (Saeman, Millett, and Lawton 1952), which may lead to progressive degradation of cellulose and production of the low-molecular-weight carbohydrates during a longer storage time. It was observed that after a storage at 22 ± 2°C for 16 weeks, the concentrations of glucose and cellobiose in 10% aqueous ethanol increased to 26 ± 11 and 92 ± 36 ppb, respectively. These high concentrations, however, were postulated as unlikely resulting from irradiation; rather they were due to the long storage time which was determined to be unrealistic for refrigerated, uncooked meat packaged with the adsorbent pad.

Colorants for Polymers

The effects of irradiation on several antioxidants and stabilizers have been studied, but colorants for polymers have not yet been studied in a systematic manner. Komolprasert and co-workers (2006) studied the effects of 10- to 20-kGy gamma radiation on two colorants present at levels of 0.1–1 wt.% in PS. The colorants were Chromophtal Yellow 2RLTS (2,3,4,5-tetrachloro-6-cyanobenzoic acid, aka Yellow 110) and Irgalite Blue GBP (copper (II) phthalocyanine blue, aka Copper II blue). Chromophtal Yellow is an organic pigment (85–90 wt.%) with 5–15 wt.% hydrogenated rosin. Irgalite Blue is an organometallic pigment (90–99 wt.%) with 1–5 wt.% polymerized rosin and 1–5 wt.% of a copper phthalocyanine derivative. Both colorants are nonvolatile, water insoluble, relatively heat stable, and are regulated under 21 CFR 178.3297 (colorants for polymers). Neither colorant is regulated for use as a component of packaging materials irradiated in contact with food.

Qualitatively, the results indicated that gamma radiation did not affect the infrared spectra of the pure colorants and the PS test specimens. Qualitative HS/GC/MS analyses identified many volatiles in the colorants, with more volatiles detected from the pure yellow colorant than the pure blue colorant, regardless of irradiation dose. Volatiles found in pure colorants were not found in PS test specimens containing colorants. Irradiation did not generate new compounds in PS containing either colorant at concentrations of up to 1 wt.%. The amounts of PS solids migrating into 10 and

50% ethanol maintained at 40°C for 10 days were in a range of 0.0035–0.013 wt.%. Total polymer dissolution of the irradiated PS test samples and HPLC/PDA analyses of the extracts showed typical residuals expected from PS (e.g., phenol, phenyl ethanol, benzaldehyde, acetophenone, and styrene) (Buchalla *et al.* 2002). Irradiation increased the concentrations of these residues, with the exception of styrene (unchanged, ca. 200 ppm). Based on GPC/PDA analysis, the extracts also contained oligomers with molecular weights <500 Da. The overall results suggested that both yellow and blue colorants are relatively stable to irradiation.

Approaches to the Safety Assessment of New Packaging Materials Irradiated in Contact with Foods

Any new packaging material not yet listed in 21 CFR 179.45 or the subject of an effective FCN or TOR exemption is subject to a pre-market safety assessment by FDA prior to irradiation in contact with food. As discussed in detail by Twaroski *et al.* (2006), the safety information required in a new submission to the Agency includes chemistry, toxicology, and environmental components. With regard to the chemistry and toxicology data, the safety assessment focuses on the likely consumer exposure and available toxicology information on those substances (i.e., packaging materials and their RPs) that become components of food from the intended use. With regard to the chemistry data, this includes the identity and amounts of migrants, as well as migration or other data, to allow the calculation of dietary exposures for the packaging materials and their RPs under the intended conditions of use. FDA recommends that all data and information be generated in accordance with the available guidance documents (http://www.cfsan.fda.gov/).

The first step to an appropriate testing protocol might involve an irradiation experiment designed to simulate the actual application conditions for determining the effects of irradiation on the packaging materials. After irradiation, the test materials are analyzed using methods and techniques that may require various analytical instruments. Because the effects of ionizing radiation on packaging material are random, the identities of potential RPs are generally unknown but, in many cases, may be deduced from the structure of the polymer or adjuvant. However, any analytical method used in the analysis for RPs in an irradiated test specimen should give some consideration to identification and quantification of an unknown migrant. GC/MS is commonly used for identification of volatile RPs, which can be achieved using a well-established chemical library database. On the other hand, LC/MS would be preferred for the identification of nonvolatile RPs. However, LC/MS is not widely available, and its use is limited due to lack of chemical databases for identifying the unknowns as well as other technical difficulties, for example, mobile phase solvents that might affect the unknown analytes. Other analytical methods and techniques such as solvent extraction and GPC may be used as well. Once the identities and levels of packaging materials and their RPs are known, exposure estimates can be determined.

Identification of RPs can depend on the test parameters, such as the analytical techniques employed or sample preparation procedures. Komolprasert and co-workers (2001) reported that headspace analyses detected formic and acetic acids in 25-kGy gamma-irradiated PET test specimens, but these acids were not detected by a thermal desorption technique. They reported that these acids possibly underwent further reactions with ethylene glycol during thermal desorption to form 1, 3-dioxolane and 2-methyl-1,3-dioxolane. Therefore, these two dioxolanes were detected at higher concentrations using thermal deosorption than by the headspace analyses. The authors also reported that acetaldehyde was detected, but was not quantifiable, and postulated that the GC column could also dictate the separation and detection of compounds and affect the identification of the unknown RPs.

In addition, different forms of test specimens can give different results. Komolprasert and co-workers (2003a) observed that the amount of volatiles detected in ground PET test specimens were greater than those detected in PET sheet specimens. For PET sheets and powders of identical mass, the larger surface area of ground PET samples contributed to faster desorption of volatiles, resulting in larger concentrations of volatiles detected. This suggests that results obtained from the powder irradiation are a worse case.

Alternatives to an Experimental Approach

An irradiation experiment is expensive and time consuming, considered that several experiments may be required in addition to the need for migration studies on irradiated materials. Hence, alternative approaches to the safety assessment are desirable. One approach is through the use of migration modeling instead of the assumption of 100% migration to food. Migration modeling based on the principles of diffusion has been accepted for determining the suitability of a recycled plastic for food contact (FDA 2006b). Therefore, a modeling approach might be considered for evaluating the safe use of packaging materials in contact with food during irradiation.

Simulation modeling of food irradiation is the subject of recent studies. Brescia and co-workers (2003) recently studied the e-beam processing of an apple and observed that the energy (dose) distribution within the apple depended on the entrance region of the electrons, the path the electrons followed as a result of scattering due to irregular (curvature) shape, and the maximum penetration distance into the apple. They reported that e-beam would not deposit uniform energy unless the apple was tilted 30° against and toward the irradiation source. The results were realized through a simulation modeling using Monte Carlo N-Particle (MCNP) software. This software is a general-purpose MCNP code that can be used for neutron, photon, electron, or coupled neutron /photon/electron transport. Their simulation results were recently validated by Kim and co-workers (2006) who used Monte Carlo simulation to calculate dose in e-beam irradiation of complex foods, using an apple as a model. They reported that the measured and calculated dose distribution values were similar. Cleland and co-workers (2005) also applied the Monte Carlo simulation to determine energy distribution in industrial X-ray food processing. They reported that the experimental measurements and the simulation results were verified for suitability and accuracy of the method between 3 and 8 MeV.

The Monte Carlo simulation may be a useful tool for predicting the energy distribution of an ionizing radiation source and the degree of its effects on packaging materials irradiated in contact with food. The dose distribution data obtained by the simulation might provide dose distribution to predict the degree of degradation in the irradiated packaging material. Theoretically, dose or energy required to degrade a polymer depends on the polymer structure and its chemical moieties or functional groups. Sadler (2004) recently applied ion chemistry to predict the amount of ionized species that may be formed in a polymer at a dose of 10 kGy. The calculations were based on the ion-pair energy of the polymer bond. Sadler concluded that this approach may be used to evaluate the levels of RPs in food. Earlier, Kothapali and Sadler (2003) irradiated EVOH samples with gamma rays at 3–10 kGy dose and analyzed the irradiated samples versus the controls. GC/MS analysis and subsequent migration testing (95% aqueous ethanol, 40°C for 10 days) suggested that irradiation did not produce any new RPs. Although acetic acid levels were found to increase after irradiation, the authors noted that acetic acid is affirmed as generally recognized as safe (GRAS) for certain uses in food under 21 CFR 184.1005 (acetic acid) and is of no safety concern. Based on the overall results, they concluded that the amounts of RPs migrating from EVOH would not result in a DC to exceed 0.5 ppb. This type of analysis may well prove to be generally useful after studies are conducted.

In addition to dietary exposure to packaging materials and their RPs, FDA's pre-market safety assessment includes an evaluation of the toxicological information on these substances. As discussed above, the identities of RPs are generally unknown but may be deduced from the structure of the polymer or adjuvant. This, in turn, might allow like RPs, such as low-molecular-weight carboxylic acids, to be grouped and evaluated as a structural class rather than individually. Bailey *et al.* (2005) recently reported on the use of structure–activity analysis (SAR) in FCN program to determine the toxicity of components of food packaging materials according to their structural similarities with many industrial chemicals that have been analyzed for toxicological concern. SAR has been shown to be a useful tool in the FCN program and has potential to be useful in the safety assessment of unknown RPs from the irradiation of packaging materials in contact with food.

Conclusions

Improved microbiological safety of food may be attained by ionizing radiation. Many packaging materials are authorized for use in the manufacture of food packaging, but only a few are authorized for use in contact with food during irradiation. FDA's safety assessment for food contact articles, including packaging materials irradiated in contact with food, hinges on likely consumer exposure to and available toxicological information on the packaging materials and their RPs produced as a result of radiation. Any testing protocols used in evaluating packaging materials incidental to irradiation should account for the possible

RPs that are generally unknown but may, in part, be deduced from the structure of the material. Monte Carlo simulation has been shown to be a useful tool in modeling the irradiation of food and may have the potential for predicting the effects of irradiation on packaging materials. FDA recently began applying the use of SAR in the safety assessment of components of packaging materials. Given that the identity of RPs are generally unknown, but may often be grouped into classes of substances, SAR may be useful. The combination of Monte Carlo simulation modeling for predicting the types of RPs as well as their amounts, in conjunction with SAR for toxicity, may prove to be useful in evaluating the safety of the packaging materials irradiated in contact with food.

References

Ahn, Dong U. and Lee, E.J. 2004. "Mechanisms and Prevention of Off-Odor Production and Color Change in Irradiated Meat." In *Irradiation of Food and Packaging: Recent Developments*, edited by Vanee Komolprasert and Kim Morehouse, pp. 43–76. ACS Symposium Series 875: Oxford Press.

Ahn, D.U. and Nam, K.C. 2004. Effects of ascorbic acid and antioxidants on color, lipid oxidation and volatiles of irradiated ground beef. *Radiation Physics and Chemistry* 71(1–2):151–156.

Andrews, L.S. and Grodner, R.M. 2004. "Ionizing Radiation of Seafood." In *Irradiation of Food and Packaging: Recent Developments,* edited by Vanee Komolprasert and Kim Morehouse, pp. 151–164. ACS Symposium Series 875: Oxford Press.

Bailey, Allan B., Chanderbhan, Ronald, Collazo-Braier, Nancy, Cheeseman, Mitchell A., and Twaroski, Michelle L. 2005. The use of structure–activity relationship analysis in the food contact notification program. *Regulatory Toxicology and Pharmacology* 42(2):225–235.

Brescia, G., Moreira, R., Brady, L., and Castell-Perez, E. 2003. Monte Carlo simulation and dose distribution of low energy electron beam irradiation of an apple. *Journal of Food Engineering* 60(1):31–39.

Buchalla, Rainer, Begley, Timothy H., and Morehouse, Kim M., 2002. Analysis of low-molecular weight radiolysis products in extracts of gamma-irradiated polymers by gas chromatography and high-performance liquid chromatography. *Radiation Physics and Chemistry* 63(3–6):837–840.

Carlsson, D.J., Krzymien, M.E., Dreschênes, L., Mercier, M., and Vachon, C. 2001. Phosphite additives and their transformation products in polyethylene packaging for γ-irradiation. *Food Additives and Contaminants* 18(6):581–591.

Cava, R., Tárrega, R., Ramirez, M.R., Mingoarranz, F.J., and Carrasco, A. 2005. Effect of irradiation on colour and lipid oxidation of dry-cured hams from free-range reared and intensively reared pigs. *Innovative Food Science & Emerging Technologies* 6(2):135–141.

Chaudry, Muhammad Ashraf, Bibi, Nizakat, Khan, Misal, Khan, Maazullah, Badshah, Amal, Qureshi, and Muhammad Jamil. 2004. Irradiation treatment of minimally processed carrots for ensuring microbiological safety. *Radiation Physics and Chemistry* 71(1–2):171–175.

Chawla, S.P. and Chander, R. 2004. Microbiological safety of shelf-stable meat products prepared by employing hurdle technology. *Food Control* 15(7):559–563.

Chuaqui-Offermanns, N. 1989. Food packaging materials and radiation processing of food: A brief review. *International Journal of Radiation Applications and Instrumentation. Part C. Radiation Physics and Chemistry* 34(6):1005–1007.

Chytiri, Stavroula, Goulas, Antonios E., Riganakos, Kyriakos A., and Kontominas, Michael G. 2006. Thermal, mechanical and permeation properties of gamma-irradiated multilayer food packaging films containing a buried layer of recycled low-density polyethylene. *Radiation Physics and Chemistry* 75(3):416–423.

Cleland, M.R., Gregoire, O., Stichelbaut, F., Gomola, I., Galloway, R.A., and Schlecht., J. 2005. Energy determination in industrial X-ray processing facilities. *Nuclear Instruments and Methods in Physics Research Section B: Beam Interactions with Materials and Atoms* 241(1–4):850–853.

Deschênes, L., Arbour, A., Brunet, F., Court, M.A., Doyon, G.D., Fortin, F., and Rodrigue, N. 1995. Irradiation of a barrier film: Analysis of some mass transfer aspects. *Radiation Physics and Chemistry* 46(4–6):805–808.

Deschênes, L., Carlsson, D.J., Wang, Y., and Labrèche, C. 2004. "Postirradiation Transformation of Additives in Irradiated HDPE Food Packaging Materials: Case Study of Irgafos 168." In *Irradiation of Food and Packaging: Recent Developments*, edited by Vanee Komolprasert and Kim Morehouse, pp. 277–289. ACS Symposium Series 875: Oxford Press.

Diehl, J.F. 1995. *Safety of Irradiated Foods*, pp. 291–293. New York: Marcel Dekker.

Ehret-Henry, J., Bouquant, J., Scholler, D., Klinck, R., and Feigenbaum, A. 1992. ¹H-NMR for the safety control of food packaging materials: analysis of extracts from polyolefin samples. *Food Additives and Contaminants* 9(4):303–314.

Fan, Xuetong, Niemira, Brendan A., and Thayer, Donald W. 2004. "Low-Dose Ionizing Radiation of Fruit Juices." In *Irradiation of Food and Packaging: Recent Developments*, edited by Vanee Komolprasert and Kim Morehouse, pp. 138–150. ACS Symposium Series 875: Oxford Press.

Farrar, Harry IV, Derr, Donald D., and Vehar, David W. 1993. Advancements in internationally accepted standards for radiation processing. *Radiation Physics and Chemistry* 42(4–6):853–856.

FDA. 1986. Irradiation in the Production, Processing, and Handling of Food. *Federal Register. Final Rule.* April 15, 1986, 51(75), 13376–13398.

FDA. 2001. Irradiation in the Production, Processing, and Handling of Food. *Federal Register. Final Rule.* February 16, 2001, 66(33), 10574–10575.

FDA. 2005. Irradiation in the Production, Processing, and Handling of Food. *Federal Register. Final Rule.* August 16, 2005, 70(157), 48057–48073.

FDA. 2006a. http://www.cfsan.fda.gov/~dms/opa-torx.html.

FDA. 2006b. http://www.cfsan.fda.gov/~dms/opa2pmnc.html#iid5.

Franz, R. and Welle, F. 2004. "Effect of Ionizing Radiation on the Migration Behavior and Sensory Properties of Plastic Packaging Materials." In *Irradiation of Food and*

Packaging: Recent Developments, edited by Vanee Komolprasert and Kim Morehouse, pp. 236–261. ACS Symposium Series 875: Oxford Press.

Goularte, L., Martins, C.G., Morales-Aizpurúa, I.C., Destro, M.T., Franco, B.D.G.M., Vizeu, D.M., Hutzler, B.W., and Landgraf, M. 2004. Combination of minimal processing and irradiation to improve the microbiological safety of lettuce (*Lactuca sativa, L.*). *Radiation Physics and Chemistry* 71(1–2):157–161.

Goulas, Antonios E., Riganakos, Kyriakos A., and Kontominas, Michael G. 2003. Effect of ionizing radiation on physicochemical and mechanical properties of commercial multilayer coextruded flexible plastics packaging materials. *Radiation Physics and Chemistry* 68(5):865–872.

Goulas, Antonios E., Riganakos, Kyriakos A., and Kontominas, Michael G. 2004a. Effect of ionizing radiation on physicochemical and mechanical properties of commercial monolayer and multilayer semirigid plastics packaging materials. *Radiation Physics and Chemistry* 69(5):411–417.

Goulas, Antonios E., Riganakos, Kyriakos A., and Kontaminas, Michael G. 2004b. "Effect of Electron Beam and Gamma Irradiation on the Migration of Plasticizers from Flexible Food Packaging Materials into Foods and Food Simulants." In *Irradiation of Food and Packaging: Recent Developments*, edited by Vanee Komolprasert and Kim Morehouse, pp. 214–235. ACS Symposium Series 875: Oxford Press.

Han, Jaejoon, Gomes-Feitosa, Carmen L., Castell-Perez, Elena, Moreira, Rosana G., and Silva, Paulo F. 2004. Quality of packaged romaine lettuce hearts exposed to low-dose electron beam irradiation. *Lebensmittel-Wissenschaft und-Technologie.* 37(7):705–715.

Javanmard, M., Rokni, N., Bokaie, S., and Shahhosseini, G. 2006. Effects of gamma irradiation and frozen storage on microbial, chemical and sensory quality of chicken meat in Iran. *Food Control* 17(6):469–473.

Jeon, Dae Hoon, Park, Gun Young, Kwak, In Shin, Lee, Kwang Ho, and Park, Hyun Jin. 2007. Antioxidants and their migration into food simulants on irradiated LLDPE film. *LWT—Food Science and Technology* 40(1):151–156.

Jo, C., Ahn, H.J., Son, J.H., Lee, J.W., and Byun, M.W. 2003. Packaging and irradiation effect on lipid oxidation, color, residual nitrite content, and nitrosamine formation in cooked pork sausage. *Food Control* 14(1):7–12.

Kawamura, Yoko. 2004. "Effects of Gamma Irradiation on Polyethylene, Polypropylene and Polystyrene." In *Irradiation of Food and Packaging: Recent Developments*, edited by Vanee Komolprasert and Kim Morehouse, pp. 262–276. ACS Symposium Series 875: Oxford Press.

Kim, Jongsoon, Rivadeneira, Ramiro G., Castell-Perez, M. Elena, and Moreira, Rosan G. 2006. Development and validation of a methodology for dose calculation in electron beam irradiation of complex shaped foods. *Journal of Food Engineering* 74(3):359–369.

Komolprasert, V. 1998. Suitability of rigid poly(polyethylene terephthalate) materials during a 25-kGy gamma irradiation. A final report prepared for Eastman Chemical Company, Shell Chemical Company, and Continental PET Technologies, Inc.

Komolprasert, V. 1999. Initial analysis of a gamma-irradiated adsorbent pad. Unpublished data.

Komolprasert, V., McNeal, Timothy.P., Agrawal, A., Adhikari, C., and Thayer, D.W. 2001. Volatile and nonvolatile compounds in irradiated semi-rigid crystalline poly (ethylene terephthalate) polymers. *Food Additives and Contaminants* 18(1):89–101.

Komolprasert, V., McNeal, T.P., Begley, T.H., and Adhikari, C. 2002. The chemical changes in an amorphous nylon polymer upon irradiation. Poster presented at the annual IFT meeting, Anaheim, CA, July 19.

Komolprasert, V., McNeal, T.P., and Begley, T.H. 2003a. Effects of gamma and electron-beam irradiation on semi-rigid amorphous polyethylene terephthalate copolymers. *Food Additives and Contaminants* 20(5):505–517.

Komolprasert, V., McNeal, T.P., and Begley, T.H. 2003b. The effects of electron beam irradiation on ethylene vinyl alcohol copolymers. Poster presented at the annual IFT meeting, Chicago IL, July 14.

Komolprasert, V., Diel, T., Sadler, G. 2006. Gamma irradiation of yellow and blue colorants in polystyrene packaging materials. *Radiation Physics and Chemistry* 75(1–2):149–160.

Kothapali, A. and Sadler, G. 2003. Determination of non-volatile radiolytic compounds in ethylene co-vinyl alcohol. *Nuclear Instruments and Methods in Physics Research Section B: Beam Interactions with Materials and Atoms* 208:340–344.

Krzymein, M.E., Carlssson, D.J., Deschênes, L., and Mercier, M. 2001. Analyses of volatile transformation products from additives in γ-irradiated polyethylene packaging. *Food Additives and Contaminants* 18(8):739–749.

Lacroix, M., Smoragiewicz, W., Jobin, M., Latreille, B., and Krzystyniak, K. 2000. Protein quality and microbiological changes in aerobically- or vacuum-packaged, irradiated fresh pork loins. *Meat Science* 56(1):31–39.

Lacroix, M., Smoragiewicz, W., Jobin, M., Latreille, B., and Krzystyniak, K. 2002. The effect of irradiation of fresh pork loins on the protein quality and microbiological changes in aerobically—or vacuum-packaged. *Radiation Physics and Chemistry* 63(3–6):317–322.

Lacroix, M. and Chiasson, F. 2004. The influence of MAP condition and active compounds on the radiosensitization of *Escherichia coli* and *Salmonella typhi* present in chicken breast. *Radiation Physics and Chemistry* 71(1–2):69–72.

Lacroix, M. and Lafortune, R. 2004. Combined effects of gamma irradiation and modified atmosphere packaging on bacterial resistance in grated carrots (*Daucus carota*). *Radiation Physics and Chemistry* 71(1–2):79–82.

Lacroix, M., Ouattara, B., Saucier, L., Giroux, M., and Smoragiewicz, W. 2004. Effect of gamma irradiation in presence of ascorbic acid on microbial composition and TBARS concentration of ground beef coated with an edible active coating. *Radiation Physics and Chemistry* 71(1–2):73–77.

Loveridge, Vicki A. and Milch, Lauren E. 2004. "Physical Evaluation of High-Dose Irradiated Multilayer Pouches." In *Irradiation of Food and Packaging: Recent Developments*, edited by Vanee Komolprasert and Kim Morehouse, pp. 314–323. ACS Symposium Series 875: Oxford Press.

Marque D., Feigenbaum, A., Dainelli, D., and Riquett, A.M. 1998. Safety evaluation of an ionized multi-layer plastic film used for vacuum cooking and meat preservation. *Food Additives and Contaminants* 5(7): 831–841.

Martins, C.G., Behrens, J.H., Destro, M.T., Franco, D.B.G.M., Vizeu, D.M., Hutzler B., and Landgraf, M. 2004. Gamma radiation in the reduction of *Salmonella spp.*

Inoculated on minimally process watercress (*Nasturtium officinalis*). *Radiation Physics and Chemistry* 71(1–2):89–93.

McNeal, T.P., Komolprasert, V., Buchalla, R., Olivo, C., and Begley, T.H. 2004. "Effects of Ionizing Radiation on Food Contact Substance." In *Irradiation of Food and Packaging: Recent Developments*, edited by Vanee Komolprasert and Kim Morehouse, pp. 214–235. ACS Symposium Series 875: Oxford Press.

Molins, R.A., Motarjemi, Y., and Käferstein, F.K. 2001. Irradiation: a critical control point in ensuring the microbiological safety of raw foods. *Food Control* 12(6): 347–356.

Montgomery, J.L., Parrish Jr., F.C., Olson, D.G., Dickson, J.S., and Niebuhr, S. 2003. Storage and packaging effects on sensory and color characteristics of ground beef. *Meat Science* 64(4):357–363.

Nam, K.C. and Ahn, D.U. 2003. Combination of aerobic packaging to control lipid oxidation and off-odor volatiles of irradiated raw turkey breast. *Meat Science* 63(3):389–395.

Paquette, Kristina E. 2004. "Irradiation of Prepackaged Food: Evaluation of the Food and Drug Administration's Regulation of the Packaging Materials." In *Irradiation of Food and Packaging: Recent Developments*, edited by Vanee Komolprasert and Kim Morehouse, pp. 182–202. ACS Symposium Series 875: Oxford Press.

Pentimalli, M., Ragni, P., Righini, G., and Capitani, D. 2000. Polymers and paper as packaging materials of irradiated food: An NMR study. *Radiation Physics and Chemistry* 57(3–6):385–388.

Pinto, P., Ribeiro, R., Sousa, L., Cabo Verde, S., Lima, M.G., Dinis, M., Santana, A., and Botelho, M.L. 2004. Sanitation of chicken eggs by ionizing radiation: functional and nutritional assessment. *Radiation Physics and Chemistry* 71(1–2):35–38.

Prakash, Anuradha and Foley, Denise 2004. "Improving Safety ad Extending Shelf Life of Fresh-Cut Fruits and Vegetables Using Irradiation." In *Irradiation of Food and Packaging: Recent Developments*, edited by Vanee Komolprasert and Kim Morehouse, pp. 90–106. ACS Symposium Series 875: Oxford Press.

Rajkowski, Kathleen T. and Fan, Xuetong 2004. "Ionizing Radiation of Seeds and Sprouts: A Review." In *Irradiation of Food and Packaging: Recent Developments*, edited by Vanee Komolprasert and Kim Morehouse, pp. 107–116. ACS Symposium Series 875: Oxford Press.

Riganakos, K.A., Koller, W.D., Ehlermann, D.A.E., Bauer, B., and Kontominas, M.G. 1999. Effects of ionizing radiation on properties of monolayer and multilayer flexible food packaging materials. *Radiation Physics and Chemistry* 54(5): 527–540.

Sadler, G.D. 2004. "Fate of Energy Absorbed by Polymers during Irradiation Treatment." In *Irradiation of Food and Packaging: Recent Developments,* edited by Vanee Komolprasert and Kim Morehouse, pp. 203–213. ACS Symposium Series 875: Oxford Press.

Saeman, J.F., Millett, M.A., and Lawton, E.J. 1952. Effect of high-energy cathode rays on cellulose. *Industrial Engineering and Chemistry* 44(12):2848–2852.

Sommers, Christopher H., Fan, Xuetong, Handel, A. Phillip, and Sokorai, Kimberly Baxendale. 2003. Effect of citric acid on the radiation resistance of *Listeria monocytogenes* and frankfurter quality factors. *Meat Science* 63(3):407–415.

Sommers, Christopher H., Keser, Natasha, Fan, Xuetong, Wallace, F. Morgan, Novak, John S., Handel, A. Philip, and Niemira, Brendan A. 2004a. "Irradiation of

Ready-to-Eat Meats: Eliminating *Listeria monocytogenes* While Maintaining Product Quality." In *Irradiation of Food and Packaging: Recent Developments*, edited by Vanee Komolprasert and Kim Morehouse, pp. 77–89. ACS Symposium Series 875: Oxford Press.

Sommers, Christopher, Fan, Xuetong, Niemira, Brandan, and Rajkowski, Kathleen. 2004b. Irradiation of ready-to-eat foods at USDA'S Eastern Regional Research Center—2003 update. *Radiation Physics and Chemistry* 71(1–2):511–514.

Stoffers, Niels H., Linssen, Josef P.H., Franz, Roland, and Welle, Frank. 2004. Migration and sensory evaluation of irradiated polymers. *Radiation Physics and Chemistry* 71(1–2):205–208.

Twaroski, Michelle L., Bartaseh, Layla I., and Bailey, Allan B. 2006. "The Regulation of Food Contact Substances in the United Sates." In *Chemical Migration and Food Contact Materials*, edited by D. Watson, K. Barnes, and R. Sinclair, pp. 17–42. Cambridge, UK: Woodhead Publishing Limited.

Welle, Frank, Mauer, A., and Franz, Roland. 2002. Migration and sensory changes of packaging materials caused by ionising radiation. *Radiation Physics and Chemistry* 63(3–6):841–844.

Zygoura, P.D., Goulas, A.E., Riganakos, K.A., and Kontominas, M.G. 2007. Migration of di-(2-ethylhexyl)adipate and acetyltributyl citrate plasticizers from food-grade PVC film into isooctane: Effect of gamma radiation. *Journal of Food Engineering* 78(3):870–877.

Appendix: Acronyms

EVA:	Ethylene-vinyl acetate
EVOH:	Ethylene-vinyl alcohol copolymer
GPC:	Gel permeation chromatography
HDPE:	High-density polyethylene
HPLC:	High performance liquid chromatography
HS/GC/MS:	Headspace/gas chromatography/mass spectrometry
LC/MS:	Liquid chromatography/mass spectrometry
LDPE:	Low-density polyethylene
LLDPE:	Linear low-density polyethylene
PA:	Polyamide (Nylon)
PDA:	Photodiode array detection
PE:	Polyethylene
PET:	Poly(ethylene terephthalate)
PP:	Polypropylene
PS:	Polystyrene
PVC:	Polyvinyl chloride
PVDC:	Poly(vinylidene chloride)

Chapter 7

RADIO FREQUENCY IDENTIFICATION SYSTEMS FOR PACKAGED FOODS

Jung H. Han, Arnold W. Hydamaka, and Yicheng Zong

Radio frequency identification (RFID) is a system that tracks items using radio waves. RFID technology is under the vast group of automatic identification (automatic ID), which includes barcode systems. RFID systems are composed of transponders (tags), readers, and computer systems. The principles of RFID systems are as follows: (1) data that are stored in tags are activated by readers when objects that have embedded tags pass through an electromagnetic zone; (2) data are then sent to a reader for decoding; and (3) decoded data are transmitted to a computer system for further processing. There are two types of RFID tags: passive and active. They differ in the use of battery, data capacities, and other properties. Frequencies of signals are also important for implementing an RFID system. RFID technologies provide potential benefits to the food industry, including supply chain management, food traceability and recall, and food safety enhancement. The advantages of this novel technology include tracing food product from farm to table at every stage of the supply chain: from production, processing, transportation, warehousing, wholesalers, to retailers, and to consumers. RFID technologies also ensure food safety and antibioterrorism. Although there are some factors to be considered, such as cost and recycling, food industries are now adopting RFID technologies, with major benefits predicted for the near future.

Introduction to RFID

RFID is a combination of wireless communication technology, information technology, and a supporting database. The contactless data transfer between the data carrier (which is a transponder or tag) and a reader makes RFID technology far more flexible than other contact identifications, such as barcode system (Finkenzeller 2003; RFID Journal Inc. 2005).

RFID is grouped under the broad category of automatic ID systems. Automatic ID system is a generic term that includes various methods of collecting, reading, and storing data nonmanually, without human errors associated with the data collection and input. Automatic ID technology identifies and captures the necessary data to be analyzed, which includes barcode system, RFID, optical character recognition, smart memory or microprocessor cards, and biometric technologies such as voice identification, fingerprinting procedures, and retinal scans.

System Architecture

A complete RFID system consists of three major components: transponder (also interchangeable with radio frequency [RF] tags), readers (made up of transceivers and antenna), and a computer system (with appropriate application software). The principle of the operation of an RFID system is as follows: (1) data are stored in tiny electronic microchips of transponders, which are embedded in or labeled on the desired objects; (2) communication between the transponders and a reader is by radio or electromagnetic waves; (3) when the transponders pass through an electromagnetic zone, the reader transmits radio wave energy with certain frequency to the transponders, which is called a reader activation signal; (4) the signal activates the transponders to send data to the reader; (5) the reader receives and decodes data; and (6) decoded data are then transmitted to the computer system for processing (Figure 7.1).

RFID System Components

RFID refers to the technology that uses devices attached to objects (packages) that transmit data to an RFID receiver. The technology involves a reusable tag, containing a microchip and an antenna. The tag is often in the form of a tiny ribbon attached to a package or can be part

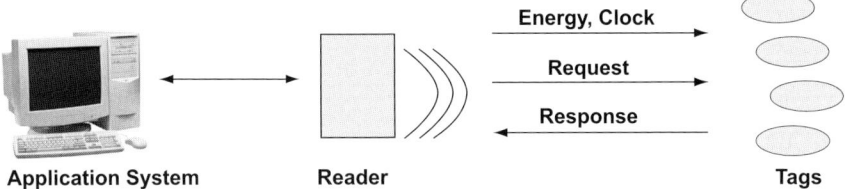

Application System   **Reader**   **Tags**

Figure 7.1. Overview of a radio frequency identification (RFID) system with passive tags (adapted from Vorst *et al*. 2004).

of the packaging material, for example, embedded between layers in a carton. RFID carries detailed information about the product in the microchip, with the added ability to collect and store data during transport and storage. RFID readers can scan and read multiple tags both rapidly and simultaneously as the tags pass in close proximity to the reader, thus rendering advantages to the industry. The information is relayed to a computer system for information gathering.

The main functions of readers are to provide energy for data communication with RFID transponders and to receive the signals from transponders (tags). Depending on different technologies, RFID readers can read data from and/or write data to the tags. They are the information control and process center for RFID systems. A reader typically consists of coupling modules, transmitter modules, control modules, and interface units. In RFID applications, readers generally communicate with readers by half-duplex for the information exchange as well as provide energy and clock to passive tags. Readers are the information carriers that realize the collection, processing, and long-range transmission of item identification through networking systems.

RFID readers generally consist of antenna, system frequency generator, coding circuit, microprocessor, memory, decoding circuit, and input–output interface. Readers can be utilized as either a handheld or a fixed-mount device. Selection of a right RFID reader usually depends on the characteristics of a model, shape and mounting requirements, dimensions of transmission zone, and the distance from next readers.

Middleware (or savant) is a software between RFID reader and commercial applications. It is an important component of an RFID system that manages readers. They take raw data from the reader, filter the data, and pass the data to backend systems (computer/database/communication systems). Middleware plays a key role in obtaining the right information to the right application at the right time.

There are many RFID middleware products on the market, but some manage RFID readers and others manage data in database-specific applications. For a specific industry, there are middleware products specially developed for its specific applications. For example, there are applications for the confirmation of shipment and receipt. The cost of middleware usually depends on the number of locations for installation and the complexity of the applications. Companies installing an RFID system will also need to purchase servers to run middleware within a warehouse, distribution center, or production facility. Even for one food product with an RFID tag from manufacturers to consumers at each stage in food channels has a different purpose to use the RFID tag. Therefore, at each stage they could use different softwares and hardwares. However, regardless of the type of software and hardware used, data obtained and processed by the middleware and servers should be shareable and communicable in a network for effective management information system.

If many tags exist in the same electromagnetic field at the same time, all tags will reply to the reader simultaneously. This will cause signal collision and indicates the presence of multiple tags. Computer system of the reader manages this collision problem by using anticollision algorithm for signal sorting and queuing.

RF protocols are developed to communicate between different transponders and a certain reader. Currently, RF protocols allows up to 1000 tags to communicate with one certain reader within a single frequency. There are two types of RF protocols: (1) tag talks first (TTF) and (2) reader talks first (RTF). The main difference between the two protocols is that RTF has a wider interference zone for other readers. Hence, RTF is very susceptible to interference from other RF devices such as cellular phones. On the other hand, TTF protocols generally work well, without interference from cellular phones, and the interference zone from other readers is much shorter (4 m). In the future, both EPCglobal and International Organization for Standardization (ISO) 18000-6 specifications will be forced to migrate to TTF-type protocols because of the high levels of interference.

RFID Tags

Tag Structure

RFID tags consist of an integrated circuit (IC), tag antenna, and a battery. A passive tag does not have battery, whereas most active tags require battery power. Tag IC contains a nonvolatile memory microchip for data storage, an AC/DC converter for processing external analog signals, internal digital data, encode/decode modulators, a logic control, and antenna connectors. This tag IC is attached to the tag antenna. The tag IC is made by silicon wafer technology; however, compared with the conventional computer chip, tag IC has a structure that is very simple and it consumes extremely low power. It converts the reader signal to supply voltage. Therefore, most tags do not require battery power. This tag chip does not have a clock generator since the chip clock is extracted from the reader signal.

The antenna is embedded in background substrate materials (i.e., paper, polyvinyl chloride [PVC], polyethylene terephthalate [PET], or other nonconductive materials). Common antenna materials are conductive silver ink, aluminum foil, or copper wire. The chip and battery unit, if required, are attached to the antenna and then covered by protective layers of polymers or paper.

Tag Frequency

Tag frequency not only decides the working principles of RFID system (magnetic coupling or electric coupling) and the reading range, but also determines the type of tag, reader, and cost. The typical RFID frequencies are 125 kHz, 133 kHz, 13.56 MHz, 27.12 MHz, 433 MHz, 902–928 MHz, 2.45 GHz, and 5.8 GHz. Among them, there are four typical frequency ranges that RFID systems run: low frequency (LF), high frequency (HF), ultra-high frequency (UHF), and microwave frequency. Generally, LF systems have short reading ranges, slow read speeds, and lower cost. HF RFID systems are utilized where longer read ranges and fast reading speeds are required. Microwave requires active RFID tags.

LF utilizes magnetic coupling working principles. The working frequency ranges are from 30 kHz to 300 kHz. The typical working frequencies are 125 kHz and 133 kHz. LF tags are normally passive tags. The reading range is shorter than 1 m. Typical applications of LF tags are animal identification, container identification, part and tool identification, and electronic antitheft keys. The working frequency ranges for HF tags are from 3 MHz to 30 MHz. The typical working frequency is 13.56 MHz. HF tags are normally passive tags. The reading range is shorter than 1 m. Typical applications of HF tags are library and electronic transit pass and electronic identifications. The working frequency ranges for UHF tags are from 433 MHz to 928 MHz. The typical working frequency is 915 MHz. UHF tags can be either passive or active tags. The reading range is normally longer than 1 m, generally 4–6 m. Working principle of UHF tags is magnetic coupling. Owing to the longer reading range, there are situations that require multiple tags for an article within the same reading range and to be read simultaneously. These days, multiple-tags reading is becoming an important characteristic of advanced RFID systems. UHF tags are usually applied in supply chain tracking by embedding in the packaging of pallets or containers. The typical working frequencies for microwave are 2.4 GHz and 5.8 GHz. Microwave tags are all active tags. The reading range can be up to 10 m. ISO 18000-6 recommends 135 kHz for LF, 13.56 MHz for HF, 860–930 MHz for UHF, and 2.45 GHz and 5.8 GHz for microwave.

Passive and Active Tags

There are two categories of RFID transponders (tags): passive and active. Passive tags are not battery supported. They do not have power sources or a transmitter. They are powered by the radio wave energy emitted by a reader. When the antenna inside the tags receives the signals produced by a magnetic field, they override data stored in chips on the signal and transmit the whole signal to the reader. Passive tags are cheaper than active tags and do not require maintenance. For these reasons, manufacturers and retailers are likely to use the passive tags in their supply chains. The reading ranges of the passive tags are much shorter than those of active tags. Passive tags are usually read-only or write-once/read-only, which means that once they have been programmed with a unique identification code, the coded data cannot be modified. The read-only passive tags cannot be rewritten or altered later.

Active tags consist of microchips mounted to an antenna and have an active transmitter on board. They are battery supported. Battery gives power to the active tags. The reading range from the RFID reader can be up to 100 m. Active tags are generally larger and more expensive than the passive tags, with more data-storage capacities, and can be read/read–write. Active tags are usually used for tracking larger assets such as railcars, large reusable containers, and cargo containers, which requires longer reading and tracking range. The cost of active tags generally ranges from $10 to $50, depending on the amount of memory on the microchip, size of the battery included, the battery life, whether onboard temperature sensor or other sensors are included, the ruggedness required, and the packaging around the transponder. Normally, active tags are not mass-produced because of the high cost. Active tags can easily detach antennas from the microchip because they are usually housed in protective plastic. Normally, a thicker and more durable plastic housing will increase the cost.

A semipassive tag has a battery but does not have a transmitter. Therefore, the semipassive tag can generate the signal without energy from the reader. However, this tag needs the electromagnetic field from readers to communicate with the readers because of the lack of transmitter. It has longer communication distance from the reader than passive tags but shorter than that of active tags.

Tag Classes

RFID is categorized into five classes by the capacity for reading and writing data. Table 7.1 summarizes the differences between and characteristics of tag classes. Class 0 is a read-only tag. This is the simplest tag containing simple identification data, which should be written by tag manufacturers. These data cannot be modified or updated later. This tag can be used for identification purposes but more widely for electronic surveillance to prevent theft and shoplifting. Class 1 is a write-once/ read-only tag that is produced by tag manufacturers as a blank memory. Data will be written by users, or tag manufacturers upon request, only once. These data cannot be erased or modified later. This class 1 tag can be used for identifying items because each tag has its own identification

Table 7.1. Differences and characteristics of tag classes.

Class	Memory	Power source	Functions	Applications
Class 0	None	Passive	Electronic surveillance system	Antitheft, identification
Class 1	Read-only	Passive or active	Electronic product code	Identification
Class 2	Read-write	Passive or active	Electronic product code	Data logging
Class 3	Read-write	Semipassive or active	Sensor tags	Wireless sensor
Class 4	Read-write	Active	Smart cards	Networking

Modified from Lewis (2004).

data, which is an electronic product code (EPC). Cost wise, passive tags without battery power are common among class 1, but active tags can also be used. Class 2 tags can read and write data multiple times. Therefore, users can add their data to update information during manufacturing and supply chain stream of products. This tag is used for data logging as well as simple identification. Class 2 tags can be used for complicated supply chain management by storing significant data at various stages of the supply chain. Class 3 tag is a read–write tag with onboard sensors. Any data collected by the sensors can be written in the tag memory. These data can include temperature, pressure, motion, or others. Because the reader can only identify data already stored in memory by the sensors and cannot correctly read the data while the sensor is recording other data, the sensor and the reader cannot work at the same time. To avoid this problem, by controlling reading and writing, semipassive or active tags are used for the sensor tags. Class 3 sensor tags have the potential to be used widely by the food industry, distributors, and retailers because of the precise control of food quality and environment. They can be used for monitoring storage temperature and vibration during distribution as well as for estimating shelf life and expiration date of foods. Class 4 tag is an integrated transmitter that communicates with other tags and devices without a reader. Owing to the power requirement of communication, an active tag with its own battery power is required. The function of the tags determines the tag price. Lower-class tags are less expensive. For RFID embedded or attached in food packaging materials, class 0 and class 1 are common tags, whereas class 1 and class 2 are useful for case and pallet identification because cases and pallets are reusable.

Communication between Tags and Readers

RFID tags and readers use electromagnetic waves for their communications. In physics, the electric field and magnetic field are not separated waves. However, owing to the characteristics of wavelength, electric and magnetic fields can be utilized by different mechanisms of communications. Long waves travel longer distance ($>100\,km$) and penetrate thick materials but have lower energy than short waves. Owing to this low energy, the signal intensity of LF tags is weak and not practical for far range. LF tags are used for near field (distance $r < \lambda/2\pi$, where λ is wavelength). The distance between LF tags and readers is very close (<1 m). With this short distance within near-field range, magnetic inductive coupling of the LF tags is the major mechanism of communication between the tag and the reader. The reader is a coil generating and receiving magnetic field. However, very high-frequency waves such as microwave or UHF have different characteristics. They are electric waves traveling with high energy but cannot penetrate barrier materials, especially materials that have high dipole momentum, such as water molecules. Because of their short wavelength and high energy, they can travel farther than long waves. They can be used for far-field backscattering mechanism for communication. Microwave and UHF tags communicate with readers using an electric field at the range of far field ($r > \lambda/2\pi$) and transmit denser signals because of their shorter wavelength. This is why UHF and microwave tags are mostly active tags and utilized for high-class (class 3 and class 4) applications with far-field range. LF and HF are mostly used for passive tags at near field. Both communication mechanisms, near-field magnetic inductive coupling and far-field electric field backscattering, are illustrated in Figure 7.2.

Response of Tags

For LF and HF that use near field, communication is achieved by inductive coupling between two coils: reader antenna and tag antenna.

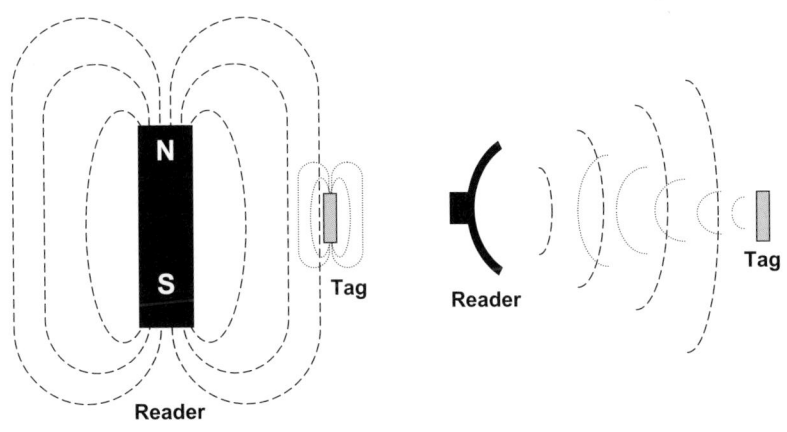

Magnetic Field Inductive Coupling   **Electric Field Back-scattering**

Figure 7.2.   Two methods of communication between reader and tag.

The reader supplies energy to tags, and the tags send information to the reader. This is a two-way radio communication. Therefore, when the reader and the tag use the same frequency, they cannot communicate simultaneously. For this reason, the tag responds with a frequency shifted from the original reader frequency. Tag information is in the side-band of the reader carrier signal, which has been modulated by the tag. This subcarrier frequency is generally a little higher than the reader carrier frequency. Since the reader and the tag transmit a slightly different frequency, they can talk simultaneously.

UHF and microwave tags use dipole momentum. This is the same fundamental mechanism of microwave ovens and surveillance radar system. At the far-field region, tags cannot be activated by inductive coupling. When the UHF and microwave from readers reach the tag antenna, the dipole energy of these waves is absorbed by the tag antenna, and part of the energy is reflected back to the reader, which is called backscattering. Instead of the side-band modulation of LF and HF, UHF and microwave use the response time shift of the backscattering signal. The time shift contains data of the tags. However, the reader can also receive the reflected signal from other objects like a radar system as well as from the tag. Therefore, the reader should have a system to eliminate noise and interference from unwanted reflection.

If tag antenna is oriented by a 90° angle to the magnetic field or dipole waves, the tag antenna cannot be coupled (in LF and HF) or pick up the dipole energy (in UHF and microwave). Therefore, to prevent this problem, multiple readers should be positioned in the field area to generate two 90° signals or in two geometric dimensions.

Frequency Allocation and Regional Regulation

RFID reader and tags are license-waived short-range devices. However, they are controlled by laws and regulations of International Telecommunication Union (ITU) and local countries. These laws and regulations vary in different countries, including the specific wavelength, purpose of the application, power emission of the electromagnetic wave, among others. ITU divides the world into three regional bodies: region 1 (European Union [EU], Africa, and former Soviet Union), region 2 (North and South America), and region 3 (Asia, Australia, and West Pacific rim). Each region has different regulations to control bands, emission power, and protocols. For example, UHF bands allowed for RFID and other wireless communications are 902–928 MHz, 869.3–869.65 MHz, and 950 MHz for regions 1, 2, and 3, respectively. Therefore, it is difficult to produce global standards or interregional standards that satisfy all these regional regulations for RFID readers and tags. Currently, the only frequency that can be accepted globally for RFID system is HF 13.56 MHz. For international trade, shipping companies or product manufacturers should attach the RFID tags to their products, which are allowed to be used in the importing final-user country. For reference information, the typical commercial microwave frequencies are 2.4 GHz and 5.8 GHz, which are currently used widely for microwave ovens, cordless telephones, and cellular phones.

EPC and Barcode

From October 1999, the AUTO-ID Labs at MIT has studied electronic tags for items and transferred its findings to EPCglobal in October 2003, which started to administer and develop EPC standards. In 2004, EPCglobal Network began developing a second-generation protocol: EPC Class 1 version 2 (also referred to as Gen 2). Gen 2 eliminates some of the incapable standards in generation 1 of AUTO-ID Labs. The goal of Gen 2 was to create a single global standard that can also be compatible with ISO standards. Gen 2 was designed to overcome the incapability between different tag protocols and overcome different regional standards. The tags will be able to work in different countries, that is, both in Europe and in Asia. End users would benefit from Gen 2, in which one international standard would be established for tracking goods through the open supply chain, especially using HF or UHF tags.

EPC Structure

The most significant function of RFID contributing to commercial industry is the EPC. EPC improves the traceability of items because of its excellent tracking capacity of each product. EPC contributes to efficient product recall and authenticity. EPC is similar to Universal Product Code (UPC), which is commonly used in barcodes. Compared with the UPC that uses 12 digits of numbers, with full combination of 10^{12}, EPC has 64 to 256 bits of alphanumeric data. The most common

EPC has 96 bits; therefore, EPC can have a full combination of 2^{96} ($=7.92 \times 10^{28}$). In reality, the first 8 bits (called header) are for the type and version of EPC, the second 28 bits (called EPC manager) are for the EPC tag manufacturer code, the third 24 bits (called object class) are for the type of a product, and the last 36 bits (called serial number) are for the coding of individual identification.

RFID and Barcodes

Barcodes identify and track products through all of the supply chain. Although efficient and convenient, barcodes have limited capacity to identify items. Single barcode only identifies one general item, not the individual object. Therefore, the same products have identical barcodes. To increase the data-storage capacity, various versions of barcodes are used, which include scattered barcode and two-dimensional barcode. Poor tracking function may be the most critical deficiency of barcode because of this limited data-coding capacity. Barcode is an optical reading system. It uses the line-of-sight technology for scanning. Therefore, scanning angle is one of the important parameters for correct data acquisition. Owing to this optical scanning method, the barcode should not be contaminated by dust, dirt, or other colored substances. The barcode should be clean and not be deformed by packages. Barcode scanning is also a very labor-intensive, slow process, and omitting and double scanning are also likely problems in supply chain management. EPC with RFID can overcome all of these barcode problems. It is an electronic system that has huge data-coding capacity in any harsh environment. Many other electronic technologies such as sensors and biosensors can be adjacent to the RFID technology, and the RFID technology can be redeveloped toward multifunctional tags beyond identification. In summary, RFID is more capable than barcode UPC because of the following: data update, data amount, reading distance, multiple reading capacity, data security, less interference to scanning, and smaller size.

Smart Label

EPC in RFID was designed to replace the UPC in barcodes. The main benefit of UPC compared with EPC is its visibility. EPC requires special readers to obtain the data; however, UPC can be read without barcode scanners. Smart labels consist of RFID, printed UPC, and other visible alphanumeric codes (and also a magnetic strip if necessary) on the surface. This label can be scanned by RFID readers and barcode scanners and also manually read and typed into computer terminals. Therefore, smart labels are human- and machine-readable EPC and UPC. These smart labels can be attached to pallets, cases, and products for identification as well as produced in a portable card form such as a credit card. A portable rigid smart card can be used as an identification/security card, bus/subway pass, e-pay toll card, credit card, or other electronic cards for public use.

EPC Applications

Data capture in computer vision now still relies on manual or semiautomatic operations. With RFID tags embedded into the target object, a computer can identify the object within 13 m from the RFID reader and capture the object and process data automatically. A wide range of machine vision applications is illustrated by the use of RFID for traffic control and vehicle access control in food processing and distribution facilities. RFID enables efficient warehousing and accurate stock control by the tracking containers, pallets, and stocktaking. RFID enables high-speed scanning of shopping baskets, trolleys, carts, stocktaking, and electronic article surveillance (EAS) in grocery stores and discount stores besides their increased convenience and accuracy of stock control. RFID technology can read the identity of many people at the same time when they pass through doorways or tube station entrances. RFID can also control the access of vehicles and other transport equipment. Throughout the supply chain of capital goods, RFID is able to read the identity of transponders mounted inside capital goods or packaging in warehouse, during transportation, and when passing through doorways. RFID technology can track case lots of low-value consumer items which are used for producers, wholesalers, and retailers. RFID tags can be embedded into containers and track the containers during shipping, air freighting, and rail movement. RFID can be used in identification, sorting, routing, and track-and-trace information for parcels, courier parcels, and documents. RFID can also be utilized for monitoring the efficiency of postal systems. With RFID, readers can identify mail while still in transit. The performance of postal systems across international boundaries can also be evaluated by using RFID (Clarke 2005). RFID can identify, track and trace, and locate documents, which is essential for insurance industry/government records. RFID can be applied to identify books, used in self-service checkout/ check-in, and can be used to identify book location.

RFID can identify, sort, and rout airline baggage. It can be used in retailing and rental of compact disc (CD) and video by identifying and stocking. RFID can identity, sort, and rout after bulk washing. RFID tags can be embedded into passports and driver licenses for identification and anticounterfeiting. RFID can be used in loyalty cards, which is convenient for remote identification of a client. It also can be used to identity and track and trace explosive items and also an antitheft measure for explosive items. RFID tags can be embedded into athletes' clothing to time marathon runners, cyclists, and other sportsmen. There are a lot of advantages that RFID technology can offer to hospitals, including tracking patients, access control, preventing baby removal, providing patient location and identification, and computerized authorization of surgical procedures. Hotels can benefit from RFID: when tags are embedded in hotel possessions, they can identify items remotely even in clients' luggage. RFID can be applied in tracking logs and products. It can also be used in inventory control of trees. RFID tag can be embedded in parts and helps to identify the parts by identification, location, time, and state. These data can help Parts and Packets Unification System to monitor, trace, and control parts and packets in constructions. RFID technology has been applied in animal identification since 1970s (Eradus and Jansen 1999). RFID is applied in animal identification and monitoring, wherein animal health and performance and reproductive status are monitored using RFID tags that have sensors (Vorst *et al.* 2004). RFID transponders have several shapes for animal identification. There are also tiny RFID transponders that can be swallowed by cows or goats and remain in the gastrointestinal tract, in order to track the feeding pattern. ISO has developed certain standards for animal identification, including ISO 11784/11785.

The goal of implementing RFID technology in pharmaceutical industries is very different from that of implementing RFID technology in other areas: in drug manufacturing, security issues, such as counterfeiting, are a huge problem worldwide. Besides, RFID can also be utilized for theft problems for popular and easily abused medications. U.S. Food and Drug Administration (FDA) supports the use of RFID technology to combat drug counterfeiting (FDA 2004, 2005).

Obstacles to RFID Applications

The first obstacle to wide utilization of RFID is the cost of tags. Tags are still expensive for use in individual item. The cost of infrastructure of RFID systems including readers, database servers, communication system with the servers, and other information technology to process huge amount of data has to be shared with all users in supply chains. The global use of EPC also requires compatibility between various regulations and standards. However, the biggest hurdle to the wide utilization of RFID for EPC is the potential problems of privacy protection. A hidden reading system could collect all data of tags of items and also RFID cardholders. Regularly, operating reading systems could be used for malicious collection of data and may be hacked for data stealing or data removing. Ethical guidelines on the use of RFID systems for data collection, data handling, and system security need to be studied intensively.

RFID for Food Industry

The implementation of RFID technology in the food industry has largely resulted from Wal-Mart in 2003 issuing a mandate to its top 100 suppliers that it would require RFID tags on all cases and pallets entering its distribution centers. Each tag would store an EPC, a new-generation barcode that could be used to track products entering the distribution centers for onward shipping. Wal-Mart has since scaled back 2005's compliance mandate, requiring its top 100 suppliers to ship tagged pallets and cases of specific stockkeeping units (SKUs) to distribution centers and stores in Texas. RFID compliance is a long-term project of Wal-Mart, which indicates that another 200 of its suppliers should be compliant by the end of 2006, with the number reaching 600 by the end of 2007. Other major players are also advocates of RFID technology, including the U.S. Department of Defense and other major stores such as Albertsons, Target, and, in Europe, Tesco and Marks & Spencer.

The major driving force to the use of RFID technology in the food industry has been for accrued benefits in greater speed and efficiency in stock rotation and better tracking of products throughout the chain, leading to improved on-shelf availability at the retail level and enhanced forecasting. Wal-Mart has had a return of investment without extensive process changes. Benefits at Wal-Mart include out-of-stock items being replenished three times faster than before, and the number of out-of-stock items that have to be manually filled has been cut by 10%, leading to additional savings in labor costs.

RFID technology has been adopted in many industries and applications; it is well suited for many operations in many types of industries, including food industries, and food supply chain management. RFID-based resource management system (RFID-RMS) can help users to handle warehouse operating orders, by retrieving and analyzing warehouse data, which can save time and cost.

Trends in Food Industry

RFID application to food packaging has certain complications that still require research and development. There are four key elements of RFID: the tag, tag reader, data monitoring/collection system, and business application software. RFID system architects are often consulted by the food company to build custom RFID solutions to the particular food package situation. Also, some consulting firms are suggesting that food companies delay implementation as long as possible, as the price of RFID tags is projected to decrease in cost. It is estimated that for food packaging, RFID tags are three to five years away from being affordable. The use of RFID in food industry is currently oriented to tracking and identification. When the RFID technology becomes more established, the integration of food science knowledge will be required to develop an intelligent food packaging application for food quality and safety (Yam, Takhistov, and Miltz 2005).

Some food companies have already integrated RFID into manufacturing. Tanirnura and Anthe Inc. and Fresh Express, both fresh-product suppliers from California, Beaver Street Fisheries, Florida, and Wells' Dairy, the largest private-label ice cream manufacturer in the United States, are examples. Implementing RFID technology at this early stage has been in response to Wal-Mart's mandate.

In addition to being suppliers to Wal-Mart, other driving forces include RFID implementation in recognition of RFID technology, potential to improve process operations and provide a competitive edge, company recognition as being responsible and a leader in the field, and, at the early stage, helping to shape technology standards and procedures. However, new technologies like RFID do not come without certain complications.

A major drawback to RFID systems that use UHF radio signals for data transmission is the interference from liquids and metals. For beverage companies and canning operations, this is a major issue. One company, KiMs, a Danish snack manufacturer, has developed innovative RFID technology by using the metal foil of the chip's packaging as an element of the tag design, overcoming the metal impediment to use of RFID (Food Quality News, 2004). Even for Wells' Dairy, ice cream tagging was an issue. Ice cream contains liquids (in a frozen form), and the frozen water behaves like a liquid in terms of the UHF radio signals by which the data are transmitted. This product interference was overcome by Wells' Dairy engineers, placing tags over an air gap in the containers. For Beaver Street Fisheries, a major technical issue resulted as to why a box of snapper is easier to read than a box of grouper. Companies with diverse product lines and having high levels of automation will likely encounter significant technical barriers to RFID implementation.

Several companies are finding that although the primary goal of RFID is tracking, process improvement, improved safety, and security are accrued benefits. This is due to the data supplied by RFID technology, as products move through the supply chain. RFID technology also provides security and safety benefits for food companies as RFID can be used to track where all packaging supplies originate from. Sea Smoke Cellars, California, winery uses RFID to track its barrels and to enhance wine making by streamlining data collection. The company plants RFID tags on tanks and bins used to harvest grapes, allowing better control of wine production and tracking.

RFID and barcodes are similar technologies in food packaging. Both systems are intended to provide rapid and reliable tracing capabilities. The primary difference between the two technologies is that barcoding is based on scanning a printed label using optical or laser systems, whereas RFID scans a tag using RF signals. Today, the major advantages of barcoding are the low cost of barcode labels and its widespread use in the food industry. However, RFID provides more traceability in food packaging than barcode labels as the tags can be scanned without direct line of sight. RFID and barcodes will both have roles in food packaging systems in the near future and offer complementary advantages.

RFID will gain continued acceptance when the benefits justify the cost and return on investment. RFID technology represents a promising food packaging innovation. By attaching an RFID tag to a package, the package becomes intelligent as the electronic chip functions as a mobile database providing valuable information that can be stored and read by appliances. This intelligent packaging technology is also being extended to refrigeration and freezing. Electrolux in Germany has demonstrated, in the food service appliance sector, the operation of nearest-expiry-first-out procedure rather than the first-in-first-out procedure as a less optimal practice.

Food packaging waste management could also benefit in the future by using RFID technology. The University of Zurich and ETH Zurich have proposed an intelligent waste management system, BIN-IT, where typical packages for consumer goods are fitted with standardized RFID tags. Waste bins or recycle centers would be equipped with RFID tag readers that would monitor volumes of certain waste types and help segregate waste packaging types, and RFID readers could award credits to members of the recycling scheme.

Supply Chain Management

The basic nature of food is that food is perishable, which is decided by the temperature, moisture content, and time. It determines the importance of food supply chain management during food distribution. At present, many foods are distributed worldwide. Thus, the most important roles of food packaging in the food supply chain are protecting and preserving foods and also promotion and reducing logistic cost (Twede 1997). RFID is a promising technology that can be applied to food supply chain management, since the major benefit of RFID technology is tracking moving items. This is mainly due to the embedded RFID transponders in pallets and other containers in food supply chain (Anon. 2005). RFID can also be applied to food supply chain management, including receiving, storage, inventories, order retrieval, shipping, retailing, and even disposing and recycling (Anon. 2005).

RFID technologies are one of the innovative technologies for food packaging applications. Placing an RFID tag to a package of food enables the supply chain to be knowledgeable about valuable information including manufacturer's history, warehouse history, locations, destinations, expiry date, temperature profiles, and even when the food is likely to be spoiled. RFID tags function like a mobile database to the food product.

Even if RFID technology is not deployed on an individual item basis, it can still provide a great deal of visibility when attached to containers carrying products throughout the supply chain. By storing such information as a product identification code and special instructions, RFID tags can make inventory management more efficient and more precise. With this technology, perishability can become less of a concern because deliveries can be received quickly, without the need for manual processing (Thomas Publishing Company 2003).

Traceability and Recall

RFID technologies also provide benefits in food recall avoidance. According to U.S. FDA, between 1999 and 2003, there were 1307 processed food product recalls in America, most of which was believed to be avoidable. The reason for processed food product recall is significant exposures to risk factors in many areas of the food supply chain. With RFID technology combined with Hazard Analysis and Critical Control Point (HACCP), an integrated and traceable supply chain can be established and managed by food processors.

Among the stages in processed food supply chain, if one of the stages breaks down, there will be a domino effect, which in turn will affect the running of the whole supply chain. All the stages in processed food supply chain—farming, cooperative processing, transportation, manufacturing, retailing, and warehousing—are equally critical to food recall problems (Stauffer 2005).

The outstanding tracing abilities of RFID tags allow a manufacturer to audit the trail of every moment of products in the retail unit and monitor correct handling, transportation, storage, and delivery. So by applying HACCP and RFID together, we can significantly reduce the number of recalls.

Food Safety and Bioterrorism

Among all the methods of food traceability, perhaps RFID is the most effective protective system. Owing to the fast and accurate tracing ability, RFID technologies can track not only items (food products) but also the individual purchase and the type of payment. RFID technologies can work with other systems such as smart labels, barcode system, tagging systems as well as management information system to enhance food safety in supply chain and food security issues such as bioterror-ism (Rasco and Bledsoe 2005).

RFID and Food Safety

Currently, food safety issues are hot. Consumers are concerned about food safety issues. In food safety matters, traceability is an important factor. RFID technology has the potential to be applied in food safety management. But there are both potential benefits and problems raised by RFID technology. RFID technology ensures high-quality data communication during supply chain and helps food industry to trace the source of the food product to provide transparency during food supply chain. RFID technology can provide a reliable connec-tion between the source of food and final food product in food supply chain, to ensure the "transparency" of the food in retail and even in restaurant and consumers' kitchen. RFID can even help food proces-sors to trace the identical animal or plant. RFID is a solution to food traceability; hence, it is an effective way of monitoring and solving food safety issues.

RFID also can address food-tampering issues. Although shrink bands and tamper-evident seals and tapes play an important role today in tamper-evidence packaging, as costs come down, RFID technology could be the next-generation tamper-evident technology. RFID tag inserted in the breakaway seal of the package, if broken could alert a monitoring system for immediate response. The connection between the tag's chip and antenna is essential for proper functioning of the tag. Similarly, cases and pallets that are RFID tagged can be readily tracked if the tagged product becomes unreadable.

The U.S. military is funding research into simple RFID sensors that could detect pathogens in food. These could be used to protect the public against food-borne illnesses or even deliberate acts of terrorism.

New Tag for Use in Metal and Metal Packaging

There are weaknesses of RFID system: when metal shielding is used, this metal affects the signal; hence, data cannot be read when tags are attached to metal surfaces and also to the inside of the packaging. High magnetic materials with less loss of radio signal are used to produce this special kind of RFID label. Up till now, these RFID tags have been used on about 10 million metal beer barrels. The working frequencies are 125–135 kHz.

Water molecules can absorb microwave signal by dipole momentum. This microwave signal absorption by water causes signal loss or interference of data acquisition from microwave RFID tags. Since most foods contain high moisture, this microwave signal interference by water should be studied thoroughly. However, this interference does not happen with LF, HF, and UHF tags.

References

Anon. 2005. What you need to know. *Transponder News*. Retrieved November 18, 2005, from http://rapidttp.com/transponder/.

Clarke, R.H. 2005. Radio frequency identifications: will it work in your supply chain? *Packexpo.com*. Retrieved November 18, 2005, from http://packexpo.com/education/msu/clarke.pdf.

FDA. 2004. Combating counterfeit drug: a report of the Food and Drug Administration. Retrieved November 18, 2005, from http://www.fda.gov/oc/initiatives/counterfeit/report02_04.html.

FDA. 2005. FDA speaks on RFID and car codes for drug packages. Ascend Media, LLC. Retrieved November 18, 2005, from http://www.fdp.com/content.php?s=FP/2004/03&p=2&sc=34.

Eradus, W.J. and Jansen, M.B. 1999. Animal identification and monitoring. *Computer and Electronics in Agriculture*. 24: 91–98.

Finkenzeller, K. 2003. *RFID Handbook*. 2nd ed. Wiley, West Sussex, UK.

Food Quality News. April 5, 2004. RFID is coming. *Food Quality News*. Retrieved November 6, 2006, from http://www.foodqualitynews.com/news/ng.asp?id=51142.

Lewis, S. 2004. *A Basic Introduction to RFID Technology and Its Use in the Supply Chain*. Laran RFID, Meyreuil, France. p. 9.

Rasco, B.A. and Bledsoe, G.E. 2005. *Bioterrorism and Food Safety*. CRC Press, Boca Raton, FL. pp. 200–221.

RFID Journal Inc. 2005. What is RFID? *RFID Journal*. Retrieved November 18, 2005, from http://www.rfidjournal.com/article/articleview/1339/1/129/.

Stauffer, J.E. 2005. Radio frequency identification. *Cereal Foods World*. 50(2): 86–87.

Thomas Publishing Company. 2003. Food packaging wrap-up. Industrial Market Trends. Retrieved November 18, 2005, from http://news.thomasnet.com/IMT/archives/2003/02/food_packaging.html.

Twede, D. 1997. Logistical/distribution packaging. In ed. Brody A. and Marsh K., *The Wiley Encyclopaedia of Packaging Technology, Second Edition*. Wiley, pp. 572–579.

Vorst, K.L., Clarke, R.H., Allison, C.P., and Booren, A.M. 2004. A research note on radio frequency transponder effects on bloom of beef muscle. *Meat Science*. 67: 179–182.

Yam, K.L., Takhistov, P.T., and Miltz, J. 2005. Intelligent packaging: concepts and applications. *Journal of Food Science*. 70(1): R1–R10.

Chapter 8

CONSUMER CHOICE: RESPONSES TO NEW PACKAGING TECHNOLOGIES

Kevin C. Spencer and Joan C. Junkus

Consumers react to packaging in a number of measurable ways. These include choice decisions to examine or ignore the packaged product, whether or not to buy it, and to buy it again. The ways that these reactions are measured are outlined. The responses of consumers to different forms and changes in food packaging have been measured and are presented. In unraveling the complexity of consumer responses to packaging attributes, and the resultant consumer choice parameters, we utilize the principles of marketing and behavioral finance. After outlining the fundamental principles underlying consumer choice, we apply these to consumer reactions to innovations, to consumer responses to packaging, and finally to consumer acceptance of novel packaging and nonthermal processing technologies. We also present original survey data of consumer responses to a variety of novel packaging and nonthermal processing technologies, in which we discover significant changes in attitudes depending upon the degree to which these technologies and their safety are explained. We conclude that there is a real need for the presentation of clear information to reduce the significant disparity between the expectations of researchers and designers who produce new packaging technologies and those of their target consumer audience.

Introduction: Customer Choice in New Packaging Technologies

A new technology can change the entire food business. Elsewhere in this volume, new packaging technologies are outlined and discussed in detail (see also Ahvenainen 2003; Han 2005). Their impact cannot yet be known, but historically, advances in packaging have created large new market segments, opened up whole new markets, and catalyzed the development of thousands of new food and beverage products (Anonymous 2003a, 2003b, 2003c). The active selection of one product over a competing product is the definition of product success. Only if a new technology is noticed and selected by consumers can it be successful in the marketplace and profitable for its manufacturers. It is crucial, then, to packaging and food manufacturers and processors to know how consumers will react and respond to these innovative technologies. If the technology is ignored or rejected, it will fail, and all of its potential benefits will remain unrealized.

In order to succeed, it is important, of course, that the new technology offers a real improvement over existing methods and is perceived by its customers to do so. But, there are two distinct groups of customers for new packaging technology, with very different reasons for deciding whether or not to buy it. The first is the food processor or manufacturer, for whom such benefits as improved handling, costs, and logistics may be very important. The second type is the end-user, the shopper who will or will not select the package because of the benefits perceived to accrue from making a choice.

Ultimately, the second type, the end-user, should be the focus of new technology efforts. It is to this shopper that both the manufactured food product and the packaging technology are aimed. It is, however, rare to market a packaging technology directly to a shopper, rather, technology is marketed to processors and manufacturers as a means to improve product sales. The shopper evaluates the new technology indirectly, based on its contribution to the quality attributes he or she desires within the food product category.

This duality must be recognized and dealt with because it is a major driver of new product success or failure. It is vital that manufacturers and food processors focus their technology efforts on the customer. How customers evaluate and make choices about new technology, what methods are used to gain insight into these decision processes, and how this information can guide designers of packaging technology are the subjects of this chapter.

The US Food Business and Current Packaging Developments

In order to intelligently discuss consumer responses to packaging, we must first briefly review the food industry and the role of packaging in it.

The US food business is enormous, with total retail sales of food and beverage in 2004 of $981.5 bn, an 8.5% increase over the year before (Standard & Poor's 2005). Of this amount, food purchases for home use totaled $529.8 bn and for eating away from home $451.7 bn (54 and 46% respectively). There were 29,365 food manufacturers reported in 2002, employing 1.52 million people (2005 data), and total food-related employment is more than 13 million people. Competition at both the manufacturing and retailing level is fierce: grocery sales in 2004 rose only by 0.9%, and the net profit of food retailers was just 1.16% (FMI 2006), down from 1.4% in 2002 (FMI 2003a). Competition for supermarket customer business in 2003 was driven by considerations of price (37%), convenience (32%), quality (26%), and variety (22%) (FMI 2003b).

Competitors distribute, present, and advertise their products in packaging. Aggressive development of new marketing designs as well as new packaging technologies represents a key factor in succeeding in the competitive marketplace. Packaging material costs represent only 8% of the retail food sector cost structure (of which 40% are paperboard), but packaging is a $79 bn industry. Growth strategies for the food industry include new product development, expanding distribution channels, and pursuing international markets, all of which require adaptations of packaging. Ongoing new product development (NPD) trends are directed at health and nutrition and ethnic specialties. For example, baby boomers, now aged 41–59 years, represent 29.4% of the population and as a consumer group focus on nutrition and weight maintenance. Food products claiming health benefits have increased to 5–10% of total product offerings. Packaging must be created which communicates health benefits, protects nutritional qualities, and meets regulatory labeling requirements. Another example of NPD focus is the ethnic market. The growth of the US Hispanic population, now 18% of the total US population, is a major driver (Standard & Poor's 2005). Packaging must be engineered to accommodate and optimize new recipe presentations and advertise them in bilingual format. A final example is the development of international markets, which currently represent a $66.8 bn share of the US processed food and beverage trade. When exported, manufacturer's brands face both opportunity and risk globally and require packaging adapted to particular markets (Buzby and Mitchell 2003).

New packaging technology plays a central role in the food industry, driving NPD, expanding major branded product lines, expanding distribution channels, facilitating export, delivering increased efficiency and lowered costs, and advertising new product benefits. These benefits accrue differently to manufacturers as opposed to end-users. As a critical component of successful NPD, new packaging technologies promise food processors and manufacturers increased product life; greater food safety; lower materials handling, and labor costs; more efficient production; and cheaper distribution. For the end-user, new technologies promise better product quality, new product types and ranges, better information exchange, and lower cost. These different benefits are emphasized in the following examples.

An example of a technology that offers to increase the efficiencies of production for food processors and manufacturers is radio frequency identification (RFID). Allowing tracking of the package and information exchange from it at every point in the distribution chain, this quantum leap improvement is now required of suppliers by the nation's largest retailers and the Department of Defense. Its implementation is expected to reduce store inventory by at least 5% and warehouse labor costs by 7.5% and will provide an unparalleled degree of food safety and security. Problems in primary (manufacturer) acceptance have included the unproven nature of efficiencies, uncertain costs, rapidly changing sizes and other manifestations of the devices employed, interference in radio transmission and receiving from various sources including packaging, and a lack of standardized approaches to use. Nevertheless, the economic benefits appear significant to manufacturers and processors, and adoption is certain. At the same time, however, the shopper is unlikely to perceive any benefit nor even is aware of the application of the technology at the point of sale (POS).

Other technologies which promise benefit to processors include ultra high pressure (UHP) processing and pulsed electric field (PEF) processing, which have been shown to inhibit spoilage microorganisms or their enzymes in the laboratory. UHP is employed on a limited basis, and PEF is in development, but providers need to demonstrate a clear cost-benefit advantage to manufacturers through better kinetic studies of microbial, spore, and enzyme inactivation, and to lower costs through optimization of process engineering (Bruin and Jongen 2003). Again, the consumer is unlikely to note these developments.

A more visible technology with clear consumer benefits for food safety, shelf life, and extended cost savings is ionizing radiation. However, there is very strong consumer opposition to the association of radiation with foods, and it is currently in limited use at only a few approved facilities in the USA (Brody 2005).

On the other hand, a development with direct appeal to consumers is active packaging (AP) technologies. Product quality indicators such as stickers, labels, or color-changing inserts respond to changes in direct (microbial growth or metabolites), indirect (temperature history, time since production, and shipment or purchase), or environmental (oxygen, carbon dioxide, or moisture content) conditions within the package. Of these, time-temperature indicators (TTI), which change color when the heat load of a refrigerated product exceeds specified limits, have advanced to commercial trials. Incorporating RFID with these measuring technologies as "intelligent packaging" (Anonymous 2003d) will allow advanced information exchange throughout the distribution system and will offer opportunities to appeal to both shoppers and manufacturers.

AP may also incorporate within the package or its constituent materials inclusions or inserts which provide moisture control, oxygen or ethylene scavenging, odor removal, antimicrobials, chemical preservatives, reactants that modify the product, edible films, or additions of desirable microflora. Of these, oxygen absorbers are commonly encountered in retail foods and represent a multimillion dollar market, but consumers are largely unaware of their presence. In fact, consumer acceptance of the broad class of AP technologies is mostly untried (Brody, Strupinsky, and Kline 2001).

Consumers' acceptance of new packaging technologies and their potential economic and social impact depend not only on their technical characteristics (Lähteenmäki and Arvola 2003) but also on the effectiveness of communication about them. Currently, US consumers know they do not like irradiation, would not be expected to care about RFID, and know little about advanced technologies such as active or intelligent packaging. This is a communication problem which is addressed variously by different providers in different markets. For example, the development of active and intelligent packaging such as TTI in Europe is inhibited by regulations on migration of substances between packaging and food, greatly limiting the permitted substances available for use in the technology. As discussed elsewhere in this volume, EU-supported efforts such as ACTIPAK are working to overcome this obstacle with a public awareness campaign stressing environmental benefits as well as shelf life and safety. Despite its untried nature, both the US and the EU markets for AP are estimated at $2 bn and are expected to have a major impact upon consumer product choices at retail.

Packaging: Function and Perception

A food package does not just enclose and protect the product, rather it sells the product (Masten 1988). The package enhances the consumers' perception of the product, increases the visibility of the product and its company, reinforces the brand image, retains current customers and attracts new ones, enhances marketing cost-effectiveness, and increases the product's competitive edge and ultimately profits. The elements important in communicating these messages are both the container itself and design elements such as graphics, color, and typeface.

Packaging is the last opportunity for a seller to trigger a consumer response, making it in essence the final salesperson (Alport 1997). While ideal shelf positioning can increase visibility by 76% (the upper shelf gets 35% more attention than the lower shelf), it is the stopping power of a package which makes the sale (Anonymous 1983). Consumers make choices very quickly (within seconds) when shopping, and a decision to further evaluate or buy is ultimately influenced most strongly by the package design (Murphy 1997). Packaging may determine a consumer's first impression of a brand, its quality or value, not only at point of purchase but also upon use. Package design is not only a key factor in product success, but innovative package design also determines brand loyalty and can be an effective counter to price competition (Carl 1995).

Assessment of the elements of packaging design which convey the impression of product quality to the consumer can be carried out by applying the theory of attractive quality which explores customer satisfaction related to varieties of perceived quality in a dynamic process (Löfgren and Witell 2005). Five classes of attributes of perceived quality have been posited and work well in categorizing consumer acceptance of changes in package design. First, attractive quality attributes are those that are unexpected but appreciated, such as a thermometer on a milk package showing the product temperature. Attractive quality attributes satisfy when present, but since they are not expected, do not dissatisfy when absent. Second, one-dimensionality attributes are those that satisfy when fulfilled, but dissatisfy when not. An example of this is a label promising 10% more milk for the same price, but actually delivering only 6%. Third, must-be qualities are those which the consumer takes for granted, but is dissatisfied when not fulfilled, for example, absence of package leakage. Fourth, indifferent qualities are those which have no impact on customer satisfaction. Reverse qualities are the final category, which are prima facie improvements, but cause dissatisfaction, such as high-tech and superior packaging that is perceived as overly complex.

The Package Design

Packaging is an essential part of selling, and the technology embodied in the package comprises a special form of information. The visual elements of packaging, such as size, shape, graphics and color, product information, and packaging technology, are key to consumer purchasing decisions (Silayoi and Speece 2004), especially in low-involvement situations, and when time is a critical factor. The impact of product imagery on consumers' beliefs about food product brands and evaluations of the brand and food product package has been studied in detail (Underwood and Klein 2002), and it is clear that consumers use packaging to infer product attributes. For instance, consumers report a more positive attitude toward the package when it includes a product picture. Product packaging, in all of its elements, is a powerful marketing communication vehicle for brand managers.

In introducing new packaging technology, the clear and consistent presentation of package elements and label must be preserved. Color and quality are important in package presentation. For instance, great effort and expense is made by graphics companies to eliminate unwanted variation (Duschene 1999). A package sits side by side on the shelf with similar units of varying age and point of manufacture, and customers will perceive any variation in label or graphic presentation as diminishment of quality.

Package Labels and Claims

The label and graphical elements of a package not only describe the product contained within, but may also make claims about the product. Consumers have very strong feelings about claims. In analyzing consumer responses toward claims made for an ecologically friendly package, Mohr, Eroğlu, and Ellen (1998) develop a valid and reliable measure of skepticism toward environmental claims. Skepticism is a cognitive response which depends upon the context and content of the communication (advertisement). A customer's skepticism to a new claim (as in a new technology) depends on experience with that particular product and brand, competitors, and the retailer, among others. This original skepticism may be strongly influenced by the customer's general attitudes about business, advertising, nutrition, or in this case, about the environment. In order to counter this original skepticism surrounding a claim for an innovation in packaging technology, convincing support for the claim must be presented in clearly and concisely worded form in the available space on the package.

Skepticism about food claims is particularly relevant to health claims. The market for foods that perform a positive health or nutrition function is large and growing rapidly, in the region of $50–$100 bn. The pitfalls of food health claims labeling and the conflict between what consumers, manufacturers, and regulators want have been well documented (Hurst 2005a) both in the USA and in the EU. Consumers are able to interpret and evaluate the FDA nutrition facts panel (Mitra *et al.* 1999), even in the presence of confusing or contradictory health benefit claims, yet respond more to advertising or media coverage of the claimed benefit, than to what the FDA label might say (Thompson 2000). General positive claims on the package are more effective than FDA-approved (and hence limited) claims for a given precise benefit. And since taste is still the most important attribute of any food product, an admixture of health and product eating quality claims can be most effective.

Consumers are motivated to buy healthy products, yet even in the burgeoning "Healthy" category, marketing the health benefits of an innovative product may have the undesired consequence of diminishing potential sales, as some consumers prefer "unhealthy" foods (Mills 2001), or because customers have become biased through experience to believe that "Healthy" products do not taste good or do not deliver on their claims. The consumer's perception of product healthiness and the resulting attitude toward the brand depends upon not only the claim made, but also upon honest disclosures of negative facts (Burton, Andrews, and Netermeyer 2000). The relationship is complex, and offering clear information is critical to consumer acceptance.

Packaging Product Pricing

While claims for ecofriendliness or healthiness attract specific segments of shoppers, all consumers are concerned with price. Providing price-per-unit information is important in helping consumers decide between products and to switch more efficiently to more cost-effective units (Granger and

Billson 1972). Consumers demand visible pricing so as to be able to compare between product choices (Langrehr and Langrehr 1983).

When not given, consumers' tendency to opt for better values is impeded.

It is very difficult for marketers to determine the optimal price for a new consumer product (Klein 1988). Business-to-business sales use a methodology based on long history and deep specialist expertise at determining pricing trade-offs, and pricing is relatively straightforward. For consumer packaged goods, on the other hand, it is not easy to establish a link between quality and price. In developing new packaging technologies, it is useful to test package attributes independently while the product is still in the design phase, but exceedingly difficult to assign price differentials to these attributes. The consumer is in fact so sensitive to pricing of product attributes in the aggregate that best assessments are always made upon completely finished products, which makes it doubly difficult for designers to value innovative attributes.

Consumer reservation prices can be estimated using a conjoint model (Jedidi and Zhang 2002) which predicts the price at which a consumer will buy a product or whether to buy at all in that product's category. The model allows simultaneous evaluation of three effects: customer switching, cannibalization, and market expansion. Such models are important in helping marketers to determine prices and pricing tactics such as bundling, target promotions, nonlinear pricing, one-to-one pricing, and assessing the impact of pricing on demand.

Impact of Package Information

Consumers depend upon price as an indicator of quality (Cronley *et al.* 2005). This is particularly true when the customer has a high need for closure (motivated to make a decision and maintain it, for instance when the choice is important monetarily or because of time pressure), when the amount of information presented is high, and when information is rank-ordered in terms of quality. Consumers must spend relatively greater effort to change closely-held beliefs in the face of new information, so that when the information load is high, there is a tendency to neglect new information and engage in selective information processing. Similarly, when information about quality attributes is rank-ordered (variable information is presented in a logical flow), consumers more easily ignore that which is inconsistent with their prior beliefs. Consumers often make purchasing judgments based on limited information or knowledge, relying on preconceived beliefs and expectations, and thus they consistently overestimate the strength of the relation between price and quality.

Inferences and choices are heavily influenced by processing ease, selective thinking, and the information value of the learning environment. Too much information on a package creates an overload which inhibits consumer purchase. Lee and Lee (2004) found that both the number of attributes and the attribute level distribution across alternatives are good predictors of overload effect. It was also shown that distant (on-line) information overload decreases all of the subjective measures (satisfaction, confusion, and confidence) compared to actual examination of the physical package. In evaluating products, consumers prefer to examine package elements and read package content, making purchase decisions with the product in hand.

In gaining acceptance of new packaging technology, it is critically important how, where, and why information is presented (the shopping environment, the package graphics elements, text, and advertising) balanced against the relative importance of the decision (cost, social pressures, and needs) and time pressures. The process is not simply logical, and the resulting choice is not the same as would be obtained from a scientific consideration of the variables. In order to decide what to put on the package, marketers need to utilize focus groups (Shapiro 1990), surveys (personal, mail, internet, and telephone), online testing (Light 2004), product trials, scanner data, sales data, manufacturing inventory, shipment, and production data. Physical methods such as eye-movement measurements (Pieters and Warlop 1999) and even functional magnetic resonance imaging (f MRI) examination have been used (Applebaum 2005) to physically and quantitatively measure the responses of consumers to new packaging. Models of consumer behavior employed depend upon the quality and completeness of the data so gathered.

What Consumers Want in a Package, Versus What Processors Want

What do shoppers want in a package? The short answer is that they want products which are easy to use and easy to store. Many studies show that convenience is the key "internal value" customers have (Tingas 2005), strongly affecting buying behavior. In a survey conducted to discover the attributes of packaged products of most concern to shoppers, Hartman, Mohan, and Mans (2005) found the six top attributes related to ease of use, five to work and clean up, three to safety, and two each to economy, health, and storage. In selecting products which deliver convenience, shoppers also look for attributes such as portability, health benefits, and great taste (Doyle 2003), demand clean and safe products and good economic value, and tamper evident packages which are easy to use, made with ecofriendly materials and fewer preservatives. In short, consumers naturally want packages to deliver everything of benefit that they can. The study determined that a high percentage of consumers would switch brands for better packaging.

Consumer preferences can change over time, which is maddening for packaging designers. For instance, reviewing the results of a 1985 survey ten years later, Spaulding (1994) noted that tamper resistance had been previously a preeminent issue because of the famous product-tampering scares at that time and that as opposed to its ubiquity later, only 60% of respondents had a microwave oven then and only 26% looked for microwaveable packaging. Also in the not-too-distant past, package safety and reliability ranked very high on shoppers list of demands (Baum 1994), along with a corollary preservation of flavor and taste (Falkman 1996), whereas both are now assumed. Main trends, however, have remained consistent. For example, convenience has always scored high (Ashton 1991), especially among time-pressed working adults, as well as among elderly buyers who like single-serve, easy-to-open, resealable containers. In addressing the current global mega trends of convenient and healthy retail and foodservice foods (Lewis 2005), processors must incorporate all of the demands of the consumer base as these evolve and accumulate over time.

Customers demand convenience plus quality, not a choice of one versus the other, and packaging is continually being developed in response to these demands (Hartnett 2001). Packaging innovation is a key factor in ensuring continued product success, while new graphics are needed to differentiate and self-sell the novel product. This is true regardless of the product category under consideration. Surveys focused upon particular product categories or assemblages yield the same trends as above, but reveal differing consumer emphases. For the frozen foods category, for example, a survey of chief shopper adults (Anonymous 2004) showed that they consider safety as a very strong attribute of frozen food packaging. Frozen boxes and bags were identified as the safest type of food packaging (39%), followed by fresh tray-wrapped (32%), dried shelf items (13%), and fully cooked and refrigerated packaged food products (11%). This holds for demographic market segments as well: elderly shoppers need packages which are easy to open, and which have print which is easy to read, while younger shoppers stress preparation convenience and time-saving features.

The dichotomy of packaging priorities between consumers and packaging manufacturers and processors is profound and is a cause of much product failure (Anonymous 2002). For example, in the foodservice category, consumer expectations of single-use takeout packaging revolve firmly around assurances of safety, tight sealing properties so as to be leak-proof and soak-proof, good insulating properties, ease of reheating, and ease of resealing after partial use. They expect labeling to clearly educate about food safety, storage, and reheating instructions (Pounder 2005). From the processors point of view, in contrast, the strongest drivers of packaging design include simplicity of operation in filling, sealing, and efficiency of materials and labor usage.

Consumers rate foodservice safety as very important (Silver 2001), and in comparison with packaged retail foods, consider takeout generally risky, and specialty meat and game packers especially risky. In contrast to consumers' particular uncertainties about game and specialty meats, processors consider them to be safe, because they are using an entirely different set of determinants of risk estimation which are unavailable to the average consumer, such as the number of years the provider has been in business (Nganje and Kaitibie 2005).

Processors have different interests than consumers, and this translates into a conflict of priorities in package design and production. For example, the top two factors for manufacturers' driving choice in purchasing packaging machinery in 2004 (Martin 2005) were compliance with new government regulations such as the Bioterrorism Act of 2002, which requires immediate source and receiver of packaged foods to be traceable, and overall production accuracy. Neither of these is of any real interest to consumers.

For processors, the top packaging trends in 2004 were (in order) product safety, cost of materials, faster line speeds, improved line automation, consumer convenience, tracking and tracing requirements, increased flexibility and changeover, new packaging materials, more customized packaging, and labeling and coding technology. These trends vary somewhat from year to year, but in general reflect the strained balance between the economic drivers of production and consumer choice. Five years ago (Higgins 2000), their top issues were (in order) consumer convenience, product safety, faster line speeds, shelf-life extension, size flexibility and customization, improved automation, flexibility and changeover, improved graphics, cost of materials, environmental concerns, and labeling and coding improvements.

Of these, only convenience and safety are really key concerns of consumers. The rest are more responsive to the manufacturing and distribution environment than to consumer choice. Processors are always focused on materials and labor costs, with automation being absolutely essential to reducing them. In new packaging development, functionality outweighs nearly everything else, including design and graphics considerations. The differences between consumers' and processors' expectations are summarized in Table 8.1.

Table 8.1.   Differences in the perceptions of packaging attributes between manufacturers and consumers.

Packaging Attributes	Manufacturer's Perception	Consumer's Perception
Production cost	Critical to cost	Irrelevant
Ease of manufacture	Critical to cost	Irrelevant
Ease of transport	Critical to cost	Irrelevant
Tamper evident	Important to liability	Expected for safety
Portion control	Important to cost	Important to value
Product advertising	Important to sell product	Important as information
Convenience	Consumer driven	Critical to choice
Price	Consumer driven	Critical to choice
Health claims	Consumer driven	Important to choice
Ease of use	Consumer driven	Important to choice
Safety	Critical	Expected
FDA/USDA certification	Required	Important to choice
Nutrition facts label	Required	Important to choice
Shelf-life	Important to logistics	Expected for value
Preservatives	Desirable	Rejected
Innovative packaging technology	Critical to growth and expansion of business	Irrelevant, except as value is demonstrated
New nonthermal processing	Critical to business growth and control of costs	Irrelevant, unless advertised

How Consumers Make Choices: Fundamentals

The fundamental framework describing how consumers make choices, whether of investments, purchases, political choices, or packaged products, is derived from basic economic theory. The basic paradigm of consumer choice assumes an always-rational consumer, equipped with complete information, and ranking all available alternatives consistently and without bias based on their ability to satisfy his/her wants. In this traditional model, a consumer chooses purchases so as to maximize his/her utility (or happiness, or satisfaction), constrained by the particular amount of money available to him to spend.

It is widely recognized that consumers do not always act this way in real life. The critical problem in analyzing individual decision-making, however, is to know and predict how people behave and will behave; that is, how people evaluate or frame a decision involving uncertainty and risk, and how psychological biases can affect that decision-making.

One rich approach to analyzing consumers' actual choice behavior studies how individuals use mental shortcuts to simplify decision-making, termed "heuristic simplification." Individuals use common mental shortcuts in order to compensate for their limited memory, less-than-perfect analytical skills, and limited time and attention to devote to choices. Studying these simplifications helps us to understand the most common simplifying strategies individuals use when sorting and evaluating information in complex situations, and the effect that these have on the decisions made.

Prospect Theory

Prospect theory applies heuristic simplification to show how a consumer frames and then evaluates, choices under uncertainty (Kahneman and Tversky 1979). The idea of analyzing the utility (or satisfaction) from a range of choices is replaced with a value function (pictured in Figure 8.1). The area between the two plotted lines represents the difference between the actual and the perceived value of the transaction to the consumer. This difference is greater for transactions which result in losses than for those which result in gains. Here, the consumer frames his/her evaluation of a purchase relative to some reference point and makes his choices in terms of gains or losses from this point. Thus, the reference point is crucial in that it anchors an individual's evaluation of a choice at a point. In addition,

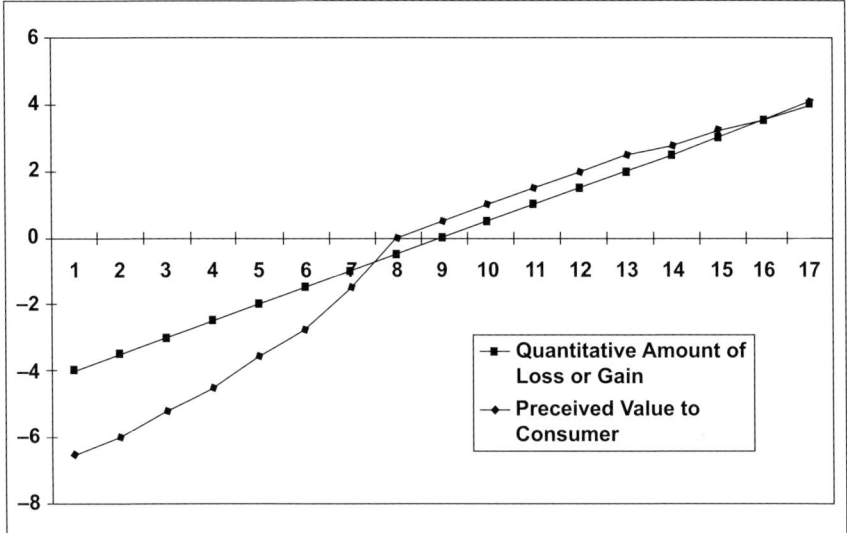

Figure 8.1.   Comparison of the actual versus perceived value of losses and gains following a consumer transaction.

the value function reflects the fact that a consumer will then focus on changes from that reference point, rather than evaluating a broad range of decisions based on absolute levels of price or other qualities of a choice.

The value function has a concave (convex) shape for gains (losses). This implies that consumers will feel happier (sadder) if a gain (loss) from a transaction is doubled, but not proportionately so: the individual will not feel himself twice as happy (or sad).

People have asymmetric reactions to losses and gains. The value function is steeper for losses than gains, and hence individuals feel worse about losing than gaining an equivalent amount. Finally, a concept related to the value function is that of mental accounting, or the tendency of the consumer to segregate gains and losses into different "mental accounts," thus separating items before evaluating their alternatives.

Reference Points

What sort of choices are made, and how an individual feels about a choice, is dependent on what is used as a reference point. Individuals generally use their current position as a base from which to make decisions going forward, and individuals generally favor a prolongation of the status quo rather than a change. This sort of behavior has been referred to as "status quo bias," or as a related "endowment effect."

The status quo bias implies that individuals will show a strong tendency to maintain their current state and will favor doing nothing rather than choose another alternative. As an example, subjects in an experiment were told that they had acquired an inheritance and were given a range of investment options in which to invest the inheritance. Subjects were more likely to choose an investment option that reflected the form of their inheritance: if the inheritance was a portfolio already invested in high-risk stocks, the subject was more likely to choose high-risk stocks as their investment option (Samuelson and Zeckhauser 1988). Thus, a consumer will often prefer to maintain a product choice rather than to choose another alternative. Further, preference for the status quo increases as the complexity of choices increases. Consumers can easily become overwhelmed by choices and cope with this complexity by doing nothing and avoiding change.

Similarly, individuals are reluctant to part with something that they already have, even if they can exchange it for something of comparable (or higher) value. This "endowment effect" arises because individuals use the status quo as their reference point and assign more weight to the loss of an object (regret) than to the gain of a new item. The endowment effect can influence purchasing decisions if some disadvantage related to a choice is framed as a loss rather than a cost (Thaler 2000). Consumers evaluate the "ordinary" price of a product as a certain amount of money, or a cost of acquiring a particular alternative. On the other hand, consumers react differently if the cost is framed as a loss. As an example, individuals were given one of two alternatives involving the cost of a theater ticket. In the first, the individual loses their $10 theater ticket as they arrive at the theater; in the second alternative, they find that they have lost a $10 bill as they arrive at the theater but before purchasing a ticket. When asked whether they would then purchase a $10 ticket, the first set of individuals were more reluctant to make the purchase, although the monetary result (losing $10 and paying $10 for a ticket) was the same in either case. The first group of individuals saw the purchaser price of the theater ticket in terms of a loss and were reluctant to purchase a ticket, while the second group perceived the price of a ticket as a cost (Kahneman and Tversky 2000).

The reference point for a particular decision can be moved or changed, a process referred to as "framing." When buying stock, an investor will generally assess the success (or not) of the investment by comparing the current value of the stock with the original purchase price. However, other prices can become the reference value. For instance, an investor will frequently assess his/her investment position at the end of the tax year, and the end of the year, the value of the stock may become the new reference point. Similarly, as the purchase price recedes in memory, the investor may instead focus on the commonly quoted 52-week high (or low) price as his new reference point (Benartzi and Thaler 1995). Selective memory is a common phenomenon: as events recede in time, individuals will tend to focus on peaks (or valleys) of their experiences, particularly if they occur at the end of an experience. For instance, an investor may feel better about an investment that experienced a dramatic gain at the end of the investment period, even if its total overall performance is less than another stock (Nofsinger 2005).

The above examples involve changes that the individual makes with regard to their reference point and their corresponding satisfaction or dissatisfaction with a choice once made. But determining the frame can influence how an individual evaluates a choice and thereby how he/she acts on it. Many choices depend critically on how the facts relevant to a decision are presented to consumers, rather than the facts themselves. For instance, a consumer will often happily pay a premium price for an ordinary item, like candy or perfume, at a vacation resort that he/she would not even purchase at a reasonable price at a regular store at home. The consumer's judgment about the price and desirability of the item is affected by the surroundings of his choice (Thaler 1985).

Gains and Losses

People pay more attention to losses than to similar gains, a tendency that contributes to the status quo effect by making individuals focus more on the disadvantages of giving up an item than the advantages of another alternative. Investors have been found to focus much more on the losses in their investment portfolio than on their gains, termed "the disposition effect." For instance, US tax laws would encourage investors to recognize losses immediately (to use the capital loss to decrease current taxes) while leaving gains alone (to minimize current taxes on their capital gains). It has been found, however, that (on average) individual investors are more likely to sell a winning stock than a losing position (Schlarbaum, Lewellen, and Lease 1978; Odean 1998). Similarly, the trading volume of stocks with large percentage price increases was significantly higher than the volume of large percentage losers (Ferris, Haugen, and Makhija 1988). Investors seem to be predisposed to recognizing gains as they occur, yet reluctant to recognize losses and experience the associated regret.

Not only do individuals have an asymmetric reaction to gains and losses, but their reaction to a gain (loss) depends on the relative size of the gain (loss) rather than its absolute size. This tendency arises because of the decreasing slope of the value function (diminishing value assigned to successive gains): twice the loss will not make an individual twice as unhappy, nor will twice the gain result in twice the happiness. This tendency is clearly at work in the popularity of consolidating credit card balances into one account even if the total owed does not change, since an individual feels relatively less worse about one big debt/loss than a sequence of smaller, yet more numerous, losses that add up to the same amount. On the other hand, sellers tend to break out and articulate all of the positive aspects of a product individually, framing a product as a collection of many, many individual attributes. Similarly, investors tend to combine their losing positions and sell them all at the same time, aggregating losses. They are less likely, however, to aggregate positive positions, and transactions involving gainers are less likely to occur together (Lim 2003).

Mental Accounting

Another very common simplifying device is to mentally segregate, label, and track different kinds of transactions and flows of money, known as "mental accounting." Individuals "label" their money by grouping expenditures into categories, separating wealth into accounts (savings versus checking), and categorizing income according to its regularity (take-home pay versus bonuses) (Shefrin and Thaler 1988). Mental accounting is used to simplify decisions, but once a label is attached to a particular money stock or flow, it is difficult to think about it in another way. Such labels may influence how individuals think about purchases and investments. For instance, investors often separate money mentally according to the object for which they are saving and treat essentially fungible funds as completely separate entities. This can lead to such behavior as borrowing at a relatively high rate to fund a consumer durable while leaving a low-rate "college fund" alone. Although they are paying more in interest than they have to, the "college" fund is labeled as off-limits financially.

Individuals also track expenses into various consumption categories and compare expenditures to mental budgets for that particular category. This is illustrated in the theater ticket example, in which the $10 ticket loss is assigned mentally to a theater account, while the $10 monetary loss is assigned to another category.

Small expenditures, however, tend not to be "booked" into specific categories but mentally labeled as "petty cash." Thus, sellers frequently frame membership fees, or charities characterize a yearly donation amount, as a small expenditure "per day" on the assumption that the expenditure will be budgeted as petty cash rather than aggregated into a budget category (entertainment and charitable donations) with some binding constraint (Gourville 1998).

People using mental accounting categories frame transactions as "closing out" particular accounts, and this can lead to a tendency to escalate commitment to a course of action (or a transaction) based on the size of "sunk costs" contrary to rational economic decision-making. A rational individual will base his decisions about future events only on present and future benefits and costs. Sunk costs, on the other hand, which are funds that have been spent no matter what the individual decides next, should have no influence on the decision to go forward. But individuals will look back to prior payments in deciding on current activities. In one experiment, season ticket holders were charged one of three levels for their season tickets, full price, or a small or large discount. The full-price group attended more events than the discounted groups, and the small discounted group more than the large discounted group (Arkes and Blumer 1985). As this example shows, the size of the sunk cost also matters: individuals will tend to show greater commitment to a choice if the sunk cost is large. The influence of a sunk cost on future decisions will diminish, however, as memories of the expense fades. This is seen, for example, in the tendency for members' usage of a health club to reach a peak right after membership dues are paid, and fall off gradually in time after this pre-payment (Gourville and Soman 1998).

Other mental heuristics include "representativeness," familiarity, and variety-seeking. Each decision-making strategy is used to simplify and/or speed up the decision process. Representativeness involves using stereotypes or simplifying relationships to classify new experiences or facts as similar to previous experiences, so that new decisions can follow the template of a previous decision. Individuals often use the most recent past as a representation of future events even if, in going forward, there is a high degree of randomness or chance. Recent mutual fund performance, or the largest gainers and losers, is reported in the financial press and used as a guide of future performance, a phenomenon known as "momentum investing" (DeBondt and Thaler 1985; DeBondt 1993). Similarly, people will tend to favor choices which are seen as more familiar to them. Employees tend to invest in their company's stock even if it would be better to diversify their holdings. Investors also tend to over-invest in firms located near to where they live. Variety-seeking can occur when a consumer must choose multiple items for consumption at various future time periods (Simonson 2000). A consumer's uncertainty about future preferences in a product category (corn flakes versus sugared cereal, chamomile tea versus Earl Grey) can be simplified and resolved by purchasing a greater variety of a product category than he or she would choose if the product were purchased and consumed sequentially. Variety packs simplify the uncertainty about future preferences and reduce the risk of future disappointment.

Finally, emotions can influence decisions. Unrelated feelings or a background mood can influence decisions involving uncertainty. Misattribution bias causes individuals to link a decision to an unrelated emotional state. Optimists, for instance, tend to underestimate risk and overestimate their chances of success, placing less weight on "bad" information and performing less critical analysis overall (Barber and Odean 2000). Even good moods can influence uncertain decisions: stock returns are significantly higher on sunny days compared to bad weather days (Hirshleifer and Shumway 2003; Kamstra, Kramer, and Levi 2003). And individuals tend to be overconfident about their ability to judge the accuracy of information and their skills in analysis. Overconfidence arises when people are given more information about a decision, even if that information does not change the probability of an accurate choice (Peterson and Pitz 1988). Individuals also are more confident in their choices if they have some measure of control over the choice: lottery players assess their chances of winning as greater if they can pick their numbers rather than being assigned a random choice (Presson and Benassi 1996).

Table 8.2.   Fundamental theory of consumer choice applied to packaging choices.

	What the Consumer Thinks About
Prospect theory	How to decide about the product
	What is the quantity of information available
	What is the weight of experience with the brand
	How strong a need there is to make this choice
Reference points	Confusion with range of varieties or attributes
	Familiarity with product
	Current favorites
	Remembered price of competing products
	Degree of novelty
Gains and losses	Anticipated cost of making a mistake
	Anticipated benefit if product performs
	Probability of failure
Mental accounting	How to decide whether to switch products
	Value for money
	Comparatively better than competition
	Comparative benefit in this category
	Degree of satisfaction with the choice

Table 8.2 summarizes the critical points of the fundamental theory of consumer choice as they apply to cognitive processing during evaluation of new packaging products. Applying these concepts to the measurement and modeling of consumers behavior as they make purchasing decisions is of great utility in guiding packaging design and marketing. In the development of new packaging technology, such knowledge allows prediction of likely choice behavior and maximizes the likelihood of product success.

How Consumers Make Choices: Measuring Decision Parameters

Packaging designers and manufacturers working with new technologies need practical information on consumer attitudes to guide them as they develop new products, and need to measure consumer choice as accurately as possible. This can be done simplistically at the store level by measuring which of several competing products is chosen, but to discover which parameters influence consumer choice and to discriminate between possible attributes to be included in a new product offering, complex statistical modeling methods are required.

In modeling consumer choice in marketing, an a priori assumption is often made that all consumers have the same preferences (aggregate discrete choice modeling). The conclusions resulting from such analyses apply to the average consumer. If consumers do not indeed vary, then these analyses are representative. However, if consumers of a category do differ significantly in their preferences, it is necessary to apply a disaggregate discrete (hierarchical Bayes) choice model (Renken 1997). The resulting data provide information about the responses of different consumer segments to new product attributes, as well as allowing the independent testing of the value to these segments of different individual product attributes.

Multidimensional scaling (MDS) methods are a powerful tool for representing (understanding) consumer preferences and choices of products and brands within a product category and are used for product positioning in marketing. Multidimensional unfolding (MDU) analyzes the individual parameters (including product attributes) of importance in complex preference/choice sets, teasing out associations within and between larger groups of products, brands, and different consumer demographics (DeSarbo, Young, and Rangaswamy 1997). These complex statistical techniques reveal strong correlations, but many have no basis in utility (are degenerate statistical artifacts) and cannot reveal real-life trends. This flaw is exacerbated for nonmetric data (qualitative or artificial preference rating scales which are commonly used in product acceptance testing) and for incomplete data (putatively representative consumer sets rather than complete populations). Using an improved statistical method which is designed to address these vulnerabilities and improve discovery of useful information, the authors present a reanalysis of data from a previous study of choices made by students and their spouses for fifteen different snack food items, and use MDU analysis to separate the preferences expressed for different products by each population, by sweetness, by consistency, and other parameters.

Conjoint analysis is a statistical modeling technique used for determining choice parameters. It assumes that a consumer's utilities remain constant during the course of measurement, but individual consumer preferences can change and evolve due to learning, fatigue, boredom, and similar factors. Liechty, Fong, and DeSarbo (2005) use dynamic Bayesian models incorporating individual heterogeneity to capture this dynamic effect, measuring variations across time as well as individuals.

This can be critical in determining the effectiveness of various informational formats (i.e., package print and media advertising) in advancing the educational process for promoting acceptance of new packaging technologies.

Similarly, in exploring the preferences that consumers have for new products, Hoeffler (2003) shows that conjoint analysis does not fully explain the inferential methods that consumers employ to evaluate "really new" versus "incrementally new" products. As consumers face much greater uncertainty in estimating the usefulness of a truly innovative product, different testing methodologies are employed which improve the predictive accuracy of preference measurement techniques. Customers may also possess multiple preferences for a given product category (Lee, Sudhir, and Steckel 2002), necessitating the use of multiple ideal point models to successfully estimate choice structure.

Such analyses can be applied not only to retail grocery shoppers, but to any customer set in any sort of transaction. Dandeo *et al.* (2004) studied the willingness of retail buyers to adopt changes in negotiation variables in automatic replenishment buying, using a model of buying behavior similar to that previously discussed for retail consumers to explore the relationships between buyers and vendors. Multiple regression analysis showed that the variables of significance in the vendor–buyer relationship, as opposed to those of concern to retail shoppers, were a merchandise-driven mentality, price/value considerations, and the color, design, and type of merchandise category. These were all strongly related to the variables common to business-to-business negotiations of price, packaging, delivery, and assortment.

Going further, relationship marketing has traditionally focused upon the relationship between suppliers and consumers. The relationship between consumers and products is equally important and less well studied. The application of acquisition pattern analysis (APA), which tends to discriminate between the structure versus the order of product acquisitions, has been undertaken (Paas, Kuijlen, and Poiesz 2005) to quantify and improve upon the brand-loyalty concept which is so important in explaining consumer–product relationships. It is upon this relationship that the producer of innovative packaging must focus.

Making Decisions by Screening

Understanding the customer choice process is of enormous importance to package designers. While the decision to purchase from a given category may be pre-determined, the decision as to which product to buy is made at the store (Stewart 2005). It is here that the shopper encounters the package, which is the representation of the product to the consumer.

Customers buy products to fulfill a need and select those that are better at doing this and are more convenient than competing products (Barwise and Meehan 2005). Further, consumers choose the brand that they expect will most reliably and best provide the generic benefits they expect from the particular product category. Extra features and aggressive branding can reinforce this but will not replace this central driver of consumer choice. The average customer expects that product quality will be sufficient to meet his/her needs and makes further choices among those products which do based on relative quality. The customer will remember those brands which deliver, as well as those which do not. In competing for customers, quality-based competitive strategies will succeed only when a company understands both those expectations and its own ability to deliver it (Hansen and Bush 1999).

Customer behavior falls short of economic rationality, since no customer uses all of the available information on every competing brand before making a choice. Rather, consumers tend to make a decision to purchase in a category, go to the store, and make a rapid brand decision based strongly on prior knowledge. In deciding between products, consumers examine several key attributes, not all of them, and therefore have limited consideration sets. Brand equity is thus very important, as familiarity with attributes simplifies decision-making. However, while the supplier cares a great deal about his brand (particularly the number and price of sales), the customer does not. All the customer cares about is the category: whether it meets his/her present need, whether it is available, and whether it is at an attractive price.

In making a product choice, most consumers screen alternatives using one or more product attributes. Alternatives which do not pass the screen will not be purchased, whereas those that do pass will be evaluated in a manner consistent with random utility theory. As shoppers, consumers encounter a multitude of product offerings and deal with complex choice sets by setting selection thresholds and utilizing a discontinuous decision process. This means that consumers first reject all choices below a certain threshold of acceptance, then evaluate those which still qualify. As an example, in choosing between packaged salads, a consumer who is strongly opposed to genetic engineering and is nervous about health risks in organic farming will first eliminate all organic products and all genetically modified organism (GMOs) (whether they are indeed so or the consumer perceives them as such) and thereafter spend effort deciding amongst the remainder (the decision set).

The first hurdle to be overcome in getting a consumer to try a new packaging product is therefore to avoid being eliminated from consideration. The consumer's discontinuous decision process utilizes screening rules. Understanding this decision process is of obvious importance to marketers, and it has been analyzed (Gilbride and Allenby 2004) as a discrete choice model accommodating conjunctive, disjunctive, and compensatory screening rules. A compensatory rule is one where the utility of the one key attribute is focused upon in evaluation and must be above the threshold value, a disjunctive rule is one where at least one attribute level among several evaluated must be acceptable and others may not be, and a conjunctive rule is one where all attributes evaluated must be acceptable. The authors find that screening rules are indeed important to consumers and that conjunctive models work best at predicting consumer behavior, with the consumers prior attitudes being very important.

Influence of Information on Decisions

The importance of prior attitudes is confirmed in studies testing the influence of information upon choice. Consumers go through two stages when considering a new product: first, whether to investigate the product more closely, and second, whether to buy it. Both of these are strongly affected by the information presented in the package design and written or imaged communication and are influenced by inferential beliefs formed by the logical processing of information from environmental and media cues. Consumers often rely upon their own generally derived beliefs and biases in making decisions about product attributes for which they have not been fully and recently educated, and rely upon sometimes faulty memory in making future purchases (Riquelme 2001). Full and immediate education, preferably at the point of purchase, is the best way to ensure that consumers' buying patterns will be the same as those expected by vendors who process mathematically derived market data, such as from scanners at POS. Consumer perceptions about product attributes are strongly influenced by external variables in their own experience and preconceptions about relations between attributes which may not be supported by the facts. If poor information is given to consumers, their inferences about product attributes will be subject to imperfect evaluation.

The amount of time and effort a consumer invests in the search, evaluation, and decision to buy a product is defined as "involvement." Involvement is a continuous variable involving multiple parameters and is difficult to quantify. Identifying the variables which are important determinants of consumer involvement has been approached through the application of fuzzy set theory (Hsu and Lee 2003), which can be used to allow marketers to measure involvement parameters quantitatively.

Examining the joint effect resulting from the interaction between product involvement and prior knowledge, Chang and Huang (2002) analyze how this impacts the utilization of different information sources by consumers in the decision-making process. When consumers have little prior knowledge of the product, external information sources will dominate the decision-making process, regardless of the level of involvement. The decision-making process is different between consumers of different levels of involvement and prior knowledge.

Influence of Brands and Brand Extensions on Decisions

Relationship marketing is the long-term interaction between suppliers and consumers, but the relationship between product and consumers is equally important (Paas, Kuijlen, and Poiesz 2005). The three critical factors determining the strength of the relationship are length of time of interaction, balance of supplier versus consumer interests, and the direction and intensity of communication. Since consumer needs and preferences change through a product life cycle, brand loyalty is a key determinant. Innovation in product packaging and technology throughout the long-term product–customer relationship, when implemented as part of a marketing strategy, can improve the overall supplier–customer relationship and increase brand loyalty.

Short-term strategies like aggressive, incentive-driven marketing may attract new customers, but they may be focused only upon the short-term offer (Dholakia 2005). In fact, high-pressure marketing tactics may discourage the brand-loyal customer. Longer term relational strategies depend upon the innovation of new packaging designs and technologies, which offer a clear set of benefits that reward the continued loyalty of present customers.

Marketers may find it difficult to succeed in brand extension through innovation because of package association barriers, where consumers associate a distinctive package type with a given product or product category (McMath 1998b). A strong relationship exists between the trust consumers have in a brand and their attitudes toward brand extensions (Reast 2005). Credibility of brand extension depends upon the effectiveness of communicating to the customer that the innovation (extension) is directly and closely related to the brand (i.e., that it is a similar product). Forays by manufacturers into sharply different categories are not likely to evoke a credible carryover response in the consumer. If a new packaging technology is to be used as a significant product attribute in a brand extension, it will best succeed if the consumer perceives that unique packaging was an important part of the original branded product. Brand loyalty comes from consistent satisfaction with product performance, and as loyalty develops, consumers spend less effort making cognitive comparisons, and just grab their favorite. Consumers who are loyal will be more likely to try brand extensions. While extension of brands is an efficient way to introduce new products, new brands can fare just as well, provided that the consumer is given sufficient useful information on the new product's features (McCarthy, Heath, and Milberg 2001).

There is a general belief among marketers that a brand that increases its product range (assortment) or variety should gain an increased market share. For example, a large consumer panel study of effect of assortment size and composition shows that adding any item improves assortment evaluation, regardless of their attributes or the size of the assortment (Oppewal and Koelemeijer 2005). However, it has been found that the type of assortment matters and can contradict this maxim (Gourville and Soman 2005). The type of assortment is defined by the degree of alignment of attributes of products within the assortment: alignable attributes vary but are readily comparable between alternatives (such as in quantitative differences of the same feature), whereas nonalignable attributes are those in which one alternative possesses the attribute and the other does not. Consumer choice is expressed as trade-offs between variations within attributes in alignable assortment types but is expressed as multidimensional choices between attributes in nonalignable assortments. The latter results in an "overchoice" effect, increasing cognitive effort and potential for regret, which has the effect of decreasing brand share. What works best in reducing or eliminating this effect is simplification of information presentation, reversibility of choice, and reduced nonalignability.

Generally, innovation increases product variety which is beneficial in meeting the diverse preferences of consumers. Consumers have a better chance of finding what they want when there are more choices and, as an individual needs or interests change, it remains more likely to find what is wanted. A retailer that offers increased selection or a manufacturer that increases its brand assortment will increase market share, and manufacturers commonly pursue this strategy, by adding new package sizes, flavors, features, formulations, and options. However, exceptions to success occur both where a retailer lowers choice but increases revenues and where increasing assortment dramatically may increase the frequency of customers not making a choice. When a customer feels overwhelmed by choice ("cognitive overload"), this strategy can backfire. Also, when customers feel higher psychological conflict about making a choice between two different but attractive alternatives, they may defer choice or remain dissatisfied with the product they have chosen, especially when choice sets are large. In order to be successful, innovations in packaging need to reinforce, not diminish, brand recognition and loyalty, and packaging must reassure the consumer about the safety and quality of the product.

How Consumers Make Choices: Acceptance of New Packaging Technology

Benefits and Risks of Innovation

Innovative food packaging and innovative processing combine to generate the high-quality, safe convenience foods which are in great demand today (Hurst 2005b). The critical elements of long shelf life and minimal preparation time are being incorporated into products with high portability and innovative designs aimed at propelling marketing. True innovation is more difficult than marginal product improvements. "New and improved" strategies are pursued by marketers who believe that consumers have notoriously short attention spans, and such products are usually perceived with great suspicion by consumers (Menzies 1998). Despite the very high levels of effort and investment made by manufacturers, most new food product innovations fail. In 2001, for example, despite a record introduction of 31,432 new domestic goods products, first-year failure rates approached 70% (ACNielsen 2001). In a German study of causes of failure, it was found that 60–75% of all innovations fail (60% in the first year and 75% by the third year) (KPMG 2003). Lack of vertical integration accounted for much of this failure rate.

Being first with a real innovation confers a significant competitive advantage, and driven by this market reality, the half-life time of NPD has decreased from around ten years in 1970 to around two years in 2000 (Bruin and Jongen 2003). Thus, process development cycles must be very fast to compete effectively. Part of the improvement in NPD is through the use of integrated process design in which concurrent engineering is employed to ensure that all phases of development are executed in parallel. This not only minimizes NPD time, but helps maintain close integration of R&D with other business functions. This is important because engineering works to create products which are feasible (highly efficient) to produce, while marketing works to create products which are appealing to consumers. In order to resolve the conflict inherent in these two opposing strategic imperatives, Michalek *et al.* (2005) propose a method, analytical target cascading, to help understand the relationship between the two and to produce the best joint optimal solution. It is built upon conjoint discrete choice modeling and demand forecasting, relating conjoint choice data to a parametric engineering design model. The result is a more profitable product.

Innovative packaging is definitely a valuable asset, as it can be and usually is a legally protected design. The design process may involve a continuing cycle of structural changes focused upon achieving and maintaining brand differential, preferably while concomitantly maintaining intellectual property rights (Buxton 2003). The design process involves ideation, modeling, and rapid prototyping and frequently results in a collision between technical practicalities and marketing design.

Bars to effective innovation include a preoccupation with cycle times that encourages spending for immediate minor improvements, while inhibiting spending on longer-term breakthroughs (Barwise and Meehan 2005; Donath 2005). Breakthrough ideas are also powerfully inhibited by excessive cost-cutting by management, obsession with short-term quarterly performance, and an aversion to financial risk inherent in the use of performance metrics like ROI, NPV, or investment payback.

A further issue complicating strategic product innovations, so critical to the marketing strategy of companies manufacturing consumer packaged goods, is that they themselves change the structure of markets (Van Heerde, Mela, and Manchanda 2004). Modeling innovation in a food product category, the authors find that introduction of a new product innovation makes existing brands seem more similar, thus decreasing brand differentiation for existing brands, and increases sales uncertainty around the time of introduction of the innovation. Encompassing changes in the dynamic food product market, the model addresses existing product heterogeneities across a category, changing parameters and nonstationarity to reveal the perturbation effect that an innovation has on an existing market. Nevertheless, linking product attributes with measured consumer choice parameters is a design which will most likely succeed.

Consumer Acceptance of Innovation

The rate of acceptance of innovative products and the pricing and marketing strategies which affect this rate are of obvious importance to vendors. Models of why consumers try new things have been the subject of much marketing science (Boyd and Mason 1999).

Consumer willingness to adopt an innovative product is termed its attractiveness, which influences consumer attitudes and intentions and the final decision to adopt (Bredahl, Grunert, and Frewer 1998). Attitude theory relates beliefs about attractiveness to expectations for the new product or technology. In adopting a true innovation, a product that is perceived as new, consumers rely upon evaluation of the attributes and characteristics of the product category rather than the brand (Boyd and Mason 1999). A product category is defined by the presence of a qualitative characteristic (e.g., calories), not a quantitative value (e.g., high or low). A market is the competitive set of products that comprise a category. All relevant category attributes are considered, including the complete market of the category: the characteristics of the individual consumer, the vendors, and those characteristics of the category which transcend the brand image.

Consumers go through a multistage decision-making process when making a purchase, as they do when evaluating innovations where they balance risk and attractiveness while comparing competing brands. The dimensions that influence adoption of innovations have been described as complexity, compatibility, observability, trialability, relative advantage, and risk. For high-tech innovations, consumers have a need for a great deal of information and support in making their choice decisions.

Mukherjee and Hoyer (2001) studied technological innovations, which are generally thought by vendors to increase product acceptance and sales. They find, however, that the positive effect of novel attributes occurs only with low-complexity products. Higher complexity, in contrast, can result in lowered product evaluation because of negative learning–cost inferences about these attributes and contribute to technophobia. Uniform standards of technology can then become important in countering consumer fears about new things.

Among models employed to explain consumer behavior (Bredahl, Grunert, and Frewer 1998) are classic multiattribute attitude models which hold that a person's attitude is determined by the sum of his/her beliefs about the consequences of, or attributes about, the product, weighted by how they are evaluated. Thus, a marketer must seek to understand all of consumers' issues with the technology globally, not just focus on the product in question. In so doing, marketers can employ refinements of this general model which posit that a consumer's intention to purchase a product is influenced by the person's attitude toward buying behavior (balance of risks and benefits), social pressure, and the degree of control a person feels they have over the buying choice.

Marketers realize that attitudes toward new innovations can change with knowledge and experience, and to accommodate this Bredahl, Grunert, and Frewer (1998) present an attitude change model which takes both attitude change and information processing theory into account. It accommodates change as it occurs both by experience (as more products are presented and used) and by information level (as knowledge is increased). The importance of the latter influence depends upon the credibility of the source, as well as on informational factors such as substantiveness of content and effective execution (degree of persuasiveness). Most manufacturers have little credibility with consumers, who expect companies to try to sell them things, and marketers need to provide information and assurances from credible sources such as government-regulatory agencies.

Role of Effective Education in Consumer Acceptance

An innovation can fail to appeal to consumers, even when the manufac-turer is certain of its value. The introduction of vacuum-packed beef did not succeed well (Eastwood 1994). Consumers thought that the color presentation was poor and maintained a preference for tray-packed meats with better color, even though vacuum-packed beef is better preserved. Producers valued the latter fact highly, as well as such producer benefits as easier and cheaper to production and shipping, lower labor costs at retail, better ease of handling, less spoilage and contamination during shipping, and a longer shelf life. Besides being driven by such internal benefits, the producers overestimated the importance to consumers of shelf life, visibility of product, and refrigerability and underestimated the perceived importance of color to the customer.

Another notable failure of a truly innovative product (McMath 1998a) was microwaveable sundaes. Processors developed the product, only to discover that consumers have no relevant relational base to heating ice cream sundaes, which is of course counterintuitive (no "chord of familiarity"). Product marketers were reluctant to spend for education, so retailers and consumers alike handled the product refrigeration suboptimally, causing poor product performance. While technical analysis confirmed the primary cause of product failure to be poor temperature control in handling, the consumer simply blamed the manufacturers for offering a product which did not perform.

As with packaging in general, consumers and processors have different views on the desirability of packaging innovations. Sherlock and Labuza (1992) studied the implementation of a specific new packaging technology, consumer TTI tags for use on refrigerated dairy products. Using focus groups and a consumer survey, it was concluded that while consumers thought it was a good idea, the success of the technology depends strongly upon both the dependability of the tag technology and the education of consumers about food spoilage.

TTI tag technology was seen by processors, in contrast, as key to the successful implementation of CAP/MAP packaging of prepared meals and other sensitive products. Refrigerated, especially dairy, products are highly sensitive to temperature change, with the rate of quality loss increasing two to eight times for each 10°C increase in temperature (the Q_{10}). Most manufacturers implement open dating to enable FIFO stock rotation. Tracking to ensure maintenance of temperature and date rotation is therefore critical, and TTI tagging can be applied to pallets or individual packages, can be scanned anywhere in distribution, and can be included as a simple consumer-readable tag (CT) which shows an OK/Not OK to the purchaser.

Results from the consumer survey indicated that about 75% of the respondents felt that both the tag and the open dating were necessary to be sure the product was safe. Most respondents felt food product safety was important, and most felt that dating alone was not sufficiently reliable. When TTI was explained carefully, 90% said that they would like to have it, would select such products, and would have more confidence in their safety. Consumers thought CT were clever, new devices but had no confidence a priori that they were reliable or could replace open dating (which is time from production to shelf), which they rely upon in the absence of clear data showing products on the shelf are actually safe or fresher. It befalls the manufacturer to instill that confidence in order to succeed with the technology.

While shoppers may find functional improvements exciting and providing added value, and may switch brands or try new products in response, they may also reject products that let them down (Doyle 2004). Acting on dissatisfaction with unmet claims, irritating and difficult-to-use packaging, and especially quality failures, consumers form strong opinions and make choices to avoid packaging products which have disappointed them. Some of these negative motivators are age related, such as the general disgust expressed by older people with the difficulty of opening blister packs. While open to exploration of novel packaging which touches upon fashion, form, function, and environmental concerns, consumers remain focused on making the best product choice easily and efficiently, and confidently obtaining the maximum product use and value. Shopper surveys have shown that consumers feel that packaging manufacturers do not have a real-life experience understanding of how their offerings impact consumers' lives.

Role of Perception of Risks and Fears in Consumer Acceptance

Consumers do not want "stuff that's bad for them," but fully expect additives and flavors to be safe (Anonymous 2005). In assessing perception of food safety risk by consumers, Mahon and Cowan (2004) find that consumers are concerned about risk across several dimensions: perceived physical risk (health endangerment), psychological risk (worry and concern about safety), and product performance risk (taste, nutrition, and value for money). The intensity of the perceived risk is dependent upon the amount at stake for the consumer. While primarily concerned with these three dimensions of risk, consumers are also concerned about three additional dimensions of risk: the time spent choosing or dealing with a product, financial costs to themselves, and the risk of social approbation.

Consumers have thresholds for each dimension of risk, and when these are crossed they will employ risk-reducing strategies: information-seeking, reliance upon brand image, consideration of price as an indicator of quality, or dependence upon a trusted retailer. These risk-relieving strategies are of varying effectiveness for different product types, and all cost the consumer time and effort. Mahon and Cowan (2004) point out that it is the perception of risk, not necessarily the actual risk that causes sales to fall for retailers and manufacturers during a scare or crisis, or in the launch of a new technology.

In choosing a product, consumers expect both safety and quality and are unwilling to trade off one for the other. They expect their perceived risk to be fully allayed by convincing informational assurances backed up by regulatory authority and by demonstrated product performance. An example of a technology which was successfully launched despite high perceived risks is the use of synthetic bovine growth hormone (rbGH) in meat production, which was made possible through dissemination of information encouraging its adoption following its approval for use by the FDA (Douthitt 1995). In dealing with real versus perceived risks, branded manufacturers must invest more in food safety to sell quality and to avoid disaster if a serious food safety breach occurs (food poisoning) which causes the name to be recognized as dangerous (Ollinger and Ballenger 2003). Consumer acceptance of any new technology or product to which they attribute some perceived risk is dependent upon their receipt of information which allays that perception.

Consumer Fears About Innovation: The Example of GMOs
There are many and manifold reasons why foods are perceived as risky, and globally, perhaps no technology is more polarizing than genetically modified foods (GMOs). Consumer fear of genetically modified food is created in large part by media stories appealing to emotional considerations (Laros and Steenkamp 2004). Consumers with strong environmental consciousness fear it more, those with good knowledge and understanding of food technology fear it less.

Perhaps no example serves better in the exploration of consumer attitudes toward risks and innovation in food product packaging than that of GMOs. In seeking to understand the difficulties inherent in the launch of GMO products and GMO-labeled packaging, it is essential both to discover current attitudes and buying behavior and to determine how these may change or be changed by producers, manufacturers, processors, and packagers. In a number of studies, Bredahl, Grunert, and Frewer (1998) show that consumers at that time generally had a low level of knowledge about genetic engineering and rejected applications of genetic technology. This negative attitude can be modeled and its determinants identified. In order to be relevant to GMOs, use of these models must incorporate general attitudes which impact acceptance of genetic engineering in food. These include perceived knowledge of genetic engineering, attitude toward environment and nature, attitude toward science and technology, food neophobia, trust in regulators, interest in food production, and price sensitivity. Vendors of GMOs need not only to provide clear assurances on packaging but to provide an education to the general public as well.

Similarly, in responding to consumer concerns about GMOs, regulators have implemented various labeling requirements. In Europe, the European Commission requires mandatory labeling for all GM food with a GM content greater than 0.9% by weight, while Canada has implemented a voluntary labeling requirement. Arguments both for and against a more stringent mandatory requirement on the EU model have been made. Public interest advocates argue that the public has a right to know the exact composition of their foods. Suppliers state that such labeling does not provide sufficient product information to permit reasoned consumer judgment, but rather results in a warning that is unintentional and unjustified.

In a climate of consumer negativity toward GM food, Canadian suppliers have no incentive to positively label food unless accompanied by a claim of benefit (health, environmental, etc.), whereas they may indeed attract customers by affixing a label stating that the product does not contain GMOs (Hu, Veeman, and Adamowicz 2005). Because of this paradox, mandatory label information is highly valued by consumers, whereas information presented voluntarily may not be. This difference in the perceived value of label information may impact the price value of the products. Under voluntary labeling, the identification of GMO content can lower the value of product, while a stated lack of GMO content may raise it.

Actual data on GMO acceptance is limited in most markets, due to the paucity of products in the market, unsettled labeling standards, and the general fear manufacturers have of advertising the fact of GMO products.

In determining current attitudes toward genetic technology in food, surveys have been undertaken (Bredahl, Grunert, and Frewer 1998) which show that while consumers had limited knowledge, they perceived GMOs to present a high risk with low benefit. These studies revealed a far lower level of acceptance of GMOs in food than in other uses (such as medicines), and lower for use in animal-derived foods than in plant-derived foods. Since consumer acceptance of food biotechnology depends upon subjective perceptions of product characteristics, especially that the ratio of benefits to risks is favorable, it is expected that acceptability to consumers will increase with increased market presence and use of products. However, consumer acceptance also depends upon perceived user benefits as opposed to producer benefits. This is confirmed in data which show that consumer resistance is focused not only upon themselves but also in concern for persons and groups of society beyond themselves (outcome groups). In considering GMOs, consumers show a conscious concern for both the product and the process which made it.

The effect of prior attitudes, source credibility, and admission of risk uncertainty by sources of information have been surveyed in Europe, where it was found that prior attitudes were most important, credibility key for those who were initially negative, and for the latter candid admissions of uncertainty by information sources created more trust. Information is key to inducing acceptance of GMOs. In the EU, where resistance is very strong, consumers demand more information which can lead to a generation of a more positive attitude when the information is presented clearly by trusted sources.

New Packaging Technologies: A Survey of Consumer Reactions

We have undertaken, as part of a much larger survey of consumer attitudes toward new food products and technologies, to assess the current state of consumer acceptance of the new AP and nonthermal processing technologies discussed in this volume. Further, we have attempted to examine the impact of information about these technologies upon consumer attitudes.

Survey Methods

Several different large groups of respondents were assembled, including "naïve" non-college educated consumers who were certainly not involved in new packaging design or distribution, and who likely had no particular familiarity with the jargon of packaging types nor with the processing methods. College-educated subgroups were also assembled, who have had arguably more technical background. The consumers sampled were presented with descriptions of technologies as they might appear on a package label and asked in a forced binary choice test to answer whether they would or not buy the product so described in a food retail shopping context. For example: "This food product has been irradiated" or "This package contains an oxygen absorber." The subjects were therefore responding to their general state of knowledge about key words in the descriptor, such as "irradiation," as well as representing the state of comprehension of the general public about such technologies as applied to food products. This would be the case of a launch of a new technology without advertising or explanatory labeling, yet with some form of simple (i.e., regulatory) labeling requirement applied.

After responding as naïve subjects, the same subjects were given a series of clear explanatory statements describing the technologies and asked once again to answer whether they would or not buy the product so described in a food retail shopping context. These statements were varied to include various representations of an assurance of the safety of the product (no mention, the word "safe," an assurance that the product had been thoroughly tested and was certified as safe by the FDA and US Government; details of its adoption and common use in familiar products, etc.). The change in these respondent attitudes was measured and compared against a similar set of respondents to whom only the second set of questions was presented. In fact, the responses of all whom answered the second set were very similar, regardless of whether they had been presented with the simpler question set previously.

Detailed analysis of the effect of inclusion of information was made for the following packaging technologies: irradiation, modified atmosphere packaging (MAP), high-pressure processing, AP, RFID, and inclusion of oxygen absorbers. Different sets of respondents (college-educated, non-college, and pooled populations) were asked whether they would buy food which was packaged with one of the technologies (no other information given), or were provided first with varying levels of information about the technology and its safety. Their responses varied depending upon the presence or absence of explanatory and reassuring information, as well as with the quantitative and qualitative degree of such information provided.

All individual paired comparison results were evaluated using chi-square, binomial, or reported probability tables. The minimum $N = 100$, although very much larger sets were generally used. Overall analyses of the larger complete survey data set were done using ANOVA and multivariate analyses to discriminate demographic and category contributory parameters, but are beyond the scope of the present article. The (alpha) $p < 0.5$ probability level was taken as significant, and numbers reported below fell within this limit.

Results: Consumer Choice of New Packaging Technologies

Our survey results reveal that most respondents feel uneasy with unfamiliar technologies and will reject them if they are offered without sufficient explanation about their composition or effect. For example, irradiation is a technology which promises great benefits, yet an enormous bias against radiation is palpable in any survey of consumer attitudes which addresses the term. As an example, in our survey, while the majority of respondents would be happy to buy packaged foods prepared which was "microwaved," less than one-quarter would accept food which had been prepared using "microwave irradiation." However, this and other negative pre-impressions can be reversed easily with a few well-chosen words of explanation and reassurance.

In the case of irradiation, only 24% of respondents initially indicated that they would accept packaged product which was irradiated, revealing a strong bias against the concept of radiation. However, acceptability became 78% when respondents were provided an informational scenario of three clear points which included statements that the process was proved safe by the US FDA, that there was absolutely no residual radiation nor other nuclear contamination present, and that it was used extensively in Europe to keep milk safe for children. For the college-educated consumer subgroup, the numbers were 35 and 85%, respectively, and for the non-college group, they were 17 and 74%, respectively.

When asked whether they would purchase foods packaged in MAP, only 50–60% of any group said yes, but that number jumped to 85–95% when the process was completely explained, they were assured that the government certifies the process as safe, and they were given examples of items they commonly purchase which are packaged in MAP. Additionally, without further explanation, less than half of respondents would buy products packaged in nitrogen, and less than a quarter would buy a product containing argon. These negative impressions disappeared when the safety and utility of these inert gases was explained as above, exactly as was found in launching argon MAP commercially (Spencer and Humphreys 2003; Spencer 2005).

In the case of RFID packaging, only 51% of non-college but 70% of college respondents would accept it without explanation, that number jumping up to 89% in the former group but only 80% in the latter with careful description of its utility and safety (two informational points). Focus group studies revealed that there was a general feeling in both groups that the RFID chips might be good for manufacturers and retailers but were not really of benefit for the consumer, who would probably bear the cost.

In questioning the acceptability of oxygen absorbers, we found that initially only slightly more than one-third of respondents would buy food which contained an oxygen scavenger, and that that number increased to only one-half of respondents when we described them (one informational point) but did not provide safety assurances. Focus groups indicated that panelists did not like the idea of things put into their food packages and, despite a technical explanation of their composition and utility, needed clear assurances of safety to change their minds.

Similarly, the change in acceptance of high-pressure treatment, which is readily understood by respondents, was rather limited but significant when even only a shallow level of information was provided—that it was safe, nothing is added to the food, and that the process works like pressure in a submarine. Negative responses were moved from 36 to 22%.

The most significant rejections were of AP. Respondents did not like the idea of antimicrobials nor chemical preservatives added to their food packaging, and only half said they would buy such products in the absence of further explanation. Considerable education is required to change this perception: when the composition of AP and its incorporation of preservatives were described, and the utility in preventing spoilage and growth of dangerous microbes stated, but no assurances of safety to the consumer were provided, 45% of respondents still rejected the technology.

The importance of provision of adequate information is summarized in Figure 8.2. Each increase in the amount and quality of information about a new technology, as discussed above, yields a significant increase

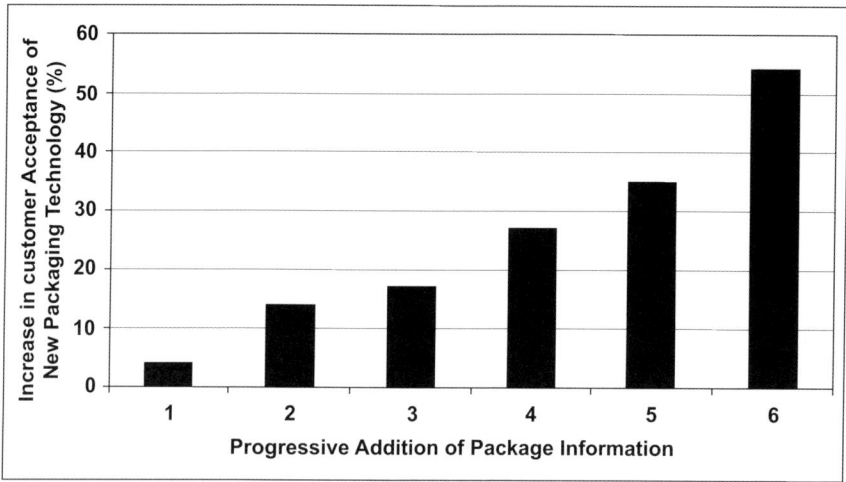

Figure 8.2.   Increase in consumer acceptance of new technologies with progressive addition of package information.

in the acceptability of that technology to the consumer. In Figure 8.2, the numbers 1–6 on the *X*-axis represent the sequential addition of information as follows: (i) a simple explanation of technology, (ii) a simple explanation with a description of its general value, (iii) a simple explanation with an assurance of safety, (iv) a complete explanation with an assurance of safety and a description of its general value, (v) a complete explanation, an assurance of safety, a certification of safety by FDA/USDA, and a discussion of use of the technology elsewhere, and (vi) a complete explanation, an assurance of safety, a certification of safety by FDA/USDA, and an example of common use in the user's experience.

These data are robust across different demographics and clearly demonstrate that consumers want to know what they are buying. Purveyors of new technologies promise benefits of which the consumer is suspicious and may include terms or descriptions which make the consumer nervous. To allay these uncertainties, consumers demand clear assurances of safety, clear explanations of what the technology is, and a clear description of the benefit which will accrue to the purchaser.

The results prove a clear need for packaging producers and distributors to inform their consumer base about new technologies, and the power of a credible representation of the safety of new food product packaging and processing technologies. They also describe a stark contrast between the stated expectations of the creators and vendors of new packaging technologies and the general consuming public.

Conclusions

Customers are very wise. They buy what they want and decide what they want for themselves. They use all of the information available to make the best choices that they can. Confronted with new packaging technology, a consumer will ask "What is it, and what's in it for me?," and the manufacturer had better have a clear, concise, compelling answer, and a product which consistently delivers on the promise of that answer at a reasonable price. Customers are well able to adapt to new technologies and are eager to find better products. New packaging technologies offer great advantages to both processors and end-users, but retail shoppers will naturally only be interested in those which are of clear benefit to themselves.

We have presented the current state of knowledge of how consumers make choices, how these mechanisms apply to choices of packaging, of innovations, and of innovative packaging. The success or failure of new packaging technologies introduced into the marketplace will depend not only upon an understanding of the parameters of consumer choice, but upon the ability of the packaging producer to deliver a product which satisfies them.

References

ACNielsen. 2001. Fifth Annual ACNielsen Consumer and Market Trends. Report. New York, New York, ACNielsen.

Ahvenainen, Raija. 2003. Editor. *Novel Food Packaging Techniques*. Boca Raton, Florida: CRC Press LLC.

Alport, H. 1997. Global, Interactive Marketing Call for Innovative Packaging. *Marketing News* 31(1): 30.

Anonymous. 1983. Packaging Research Probes Stopping Power, Label Reading, and Consumer Attitudes among the Targeted Audience. *Marketing News* 17(15): 8.

Anonymous. 2002. Packaging Report Card. *BrandPackaging* 6(9): 16.

Anonymous. 2003a. Packages: Tracing an Evolution. *Packaging Digest* (Dec): 37–42.

Anonymous. 2003b. Equipment in a Starring Role. *Packaging Digest* (Dec): 45–56.

Anonymous. 2003c. Materials Shape the Future of Packaging. *Packaging Digest* (Dec): 59–63.

Anonymous. 2003d. When Talking Foods Start Walking: The Future of Active and Intelligent Packaging. *Food Engineering & Ingredients* 28(1): 38.

Anonymous. 2004. Study Examines Marketing, Channel and Packaging Attitudes. *Frozen Food Age* 52(11): 45.

Anonymous. 2005. Insight-Food Additives: A Truly Acquired Taste. *Marketing Week* (Sep 1): 26–27.

Applebaum, M. 2005. Neuroscience: MRI to ROI. *Brandweek* 46(10): 34.

Arkes, H.R., and C. Blumer. 1985. The Psychology of Sunk Cost. *Organizational Behavior and Human Decision Processes* 35(1): 124–140.

Ashton, R. 1991. Consumers Want it All. *Packaging* 36(7): 32–36.

Barber, B., and T. Odean. 2000. Trading is Hazardous to Your Wealth: The Common Stock Investment Performance of Individual Investors. *Journal of Finance* 55: 773–806.

Barwise, P., and S. Meehan. 2005. Simply Better. *Marketing Research* 17(2): 9–14.

Baum, C. 1994. 10th Annual Packaging Consumer Survey 1994: Consumers Want it All-and Now. *Packaging* 39(8): 40–43.

Benartzi, S., and R. Thaler. 1995. Myopic Loss-Aversion and the Equity Premium Puzzle. *The Quarterly Journal of Economics* 110: 75–92.

Boyd, T.C., and C.H. Mason. 1999. The Link Between Attractiveness of "Extrabrand" Attributes and the Adoption of Innovations. *Journal of the Academy of Marketing Science* 27(3): 306–319.

Bredahl, L., K.G. Grunert, and L.J. Frewer. 1998. Consumer Attitudes and Decision-Making with Regard to Genetically Engineered Food Products—A Review of the Literature and a Presentation of Models for Future Research. *Journal of Consumer Policy* 21: 251–277.

Brody, A.L. 2005. Packaging for Nonthermally & Minimally Processed Foods. *Food Technology* 59(10): 75–77.

Brody, A.L., E.R. Strupinsky, and L.R. Kline. 2001. *Active Packaging for Food Applications*. Lancaster, Pennsylvania: Technomic Publishing Company, Inc.

Bruin, S., and Th.R.G. Jongen. 2003. Food Process Engineering: The Last 25 Years and Challenges Ahead. *Comprehensive Reviews in Food Science and Food Safety* 2: 42–81.

Burton, S., J.C. Andrews, and R.G. Netermeyer. 2000. Nutrition Ad Claims and Disclosures: Interaction and Mediation Effects for Consumer Evaluations of the Brand and the Ad. *Marketing Letters* 11(3): 235–247.

Buxton, P. 2003. Why Shape Matters. *Marketing* (Nov 13): 31–32.

Buzby, J.C., and L. Mitchell. 2003. Food Safety and Trade: Regulations, Risks, and Reconciliation. *Amber Waves* (Nov) USDA-ERS.

Carl, K. 1995. Good Package Design Helps Increase Consumer Loyalty. *Marketing News* 29(13): 4.

Chang, C.-H., and C.-W. Huang. 2002. The Joint Effect of Product Involvement and Prior Knowledge on the Use of Information Sources and the Choice of Decision-Making Paths by Consumers. *International Journal of Management* 19(2): 315–322.

Cronley, M.L., S.S. Posavac, T. Meyer, F.R. Kardes, and J.J. Kellaris. 2005. A Selective Hypothesis Testing Perspective on Price-Quality Inference and Inference-Based Choice. *Journal of Consumer Psychology* 15(2): 159–169.

Dandeo, L.M., S.S. Fiorito, L. Guinipero, and D.H. Pearcy. 2004. Determining Retail Buyers' Negotiation Willingness for Automatic Replenishment Programs. *Journal of Fashion Marketing & Management* 8(1): 27–40.

DeBondt, W. 1993. Betting on Trends: Intuitive Forecasts of Financial Risk and Return. *International Journal of Forecasting* 9: 355–371.

DeBondt, W., and R. Thaler. 1985. Does the Stock Market Overreact? *Journal of Finance* 40: 793–808.

DeSarbo, W.S., M. R. Young, and A. Rangaswamy. 1997. A Parametric Multidimensional Unfolding Procedure for Incomplete Nonmetric Preference/Choice Set Data in Marketing Research. *Journal of Marketing Research* 34(4): 499–516.

Dholakia, P.M. 2005. The Hazards of Hounding. *Harvard Business Review* 83(10): 20, 24.

Donath, B. 2005. Marketers: Rethink Innovation Approach. *Marketing News* 39(18): 6–8.

Douthitt, R.A. 1995. Consumer Risk Perception and Recombinant Bovine Growth Hormone: The Case for Labeling Dairy Products Made from Untreated Herd Milk. *Journal of Public Policy & Marketing* 14(2): 328–330.

Doyle, M. 2003. What Do Shoppers Want in a Package? *Food & Drug Packaging* 67(8): 16.

Doyle, M. 2004. Consumers Have Long List of Packaging Wishes and Pet Peeves: New Report Card Focuses on What Makes Packages Sell or Sit on the Shelf. *Food & Drug Packaging* 68(8): 24.

Duschene, S. 1999. Multicolor Fleet Ensures Consistency. *Graphics Arts Monthly* 71(6): 18.

Eastwood, D.B. 1994. Consumer Acceptance of New Experience Good: A Case Study of Vacuum Packed Fresh Beef. *Journal of Consumer Affairs* 28(2): 300–312.

Falkman, M.A. 1996. Consumers Buy Packaging Benefits: A Packaging Digest Exclusive Survey. *Packaging Digest* 33(10): 24–25.

Ferris, S.P., R.A. Haugen, and A.K. Makhija. 1988. Predicting Contemporary Volume with Historic Volume at Differential Price Levels: Evidence Supporting the Disposition Effect. *Journal of Finance* 43: 677–697.

FMI. 2003a. *The State of the Food Retail Industry. Food Marketing Industry Speaks.* Washington, D.C., Food Marketing Institute.

FMI. 2003b. *Trends in the United States: Consumer Attitudes & The Supermarket.* Washington, D.C., Food Marketing Institute.

FMI. 2006. *Supermarket Facts. Industry Overview 2004.* Washington, D.C., Food Marketing Institute.

Gilbride, T.J., and G.M. Allenby. 2004. A Choice Model with Conjunctive, Disjunctive, and Compensatory Screening Rules. *Marketing Science* 23(3): 391–406.

Gourville, J.T. 1998. Pennies a Day: The Effect of Temporal Reframing on Transaction Evaluation. *Journal of Consumer Research* 24(3): 395–408.

Gourville, J.T., and D. Soman. 1998. Payment Depreciation: The Effects of Temporally Separating Payments from Consumption. *Journal of Consumer Research* 25(2): 160–174.

Gourville, J.T., and D. Soman. 2005. Overchoice and Assortment Type: When and Why Variety Backfires. *Marketing Science* 24(3): 382–395.

Granger, C.W.J., and A. Billson. 1972. Consumers Attitudes Toward Package Size and Price. *Journal of Marketing Research* 9(3): 239–248.

Han, J. 2005. Editor. *Innovations in Food Packaging.* San Diego, California: Elsevier Academic Press.

Hansen, E., and R.J. Bush. 1999. Understanding Customer Quality Requirements: Model and Application. *Industrial Marketing Management* 28(2): 119–130.

Hartman, L.R., A.M. Mohan, and J. Mans. 2005. Portability, Convenience Reign Supreme in Food Finds. *Packaging Digest* 42(7): 30–39.

Hartnett, M. 2001. Innovations Open Up New Capabilities. *Frozen Food Age* 50(3): 56, 64.

Higgins, K.T. 2000. 15th Annual Packaging Trends Survey. *Food Engineering* 72(11): 46–54.

Hirshleifer, D., and T. Shumway. 2003. Good Day Sunshine: Stock Returns and the Weather. *Journal of Finance* 58: 1009–1032.

Hoeffler, S. 2003. Measuring Preferences for Really New Products. *Journal of Marketing Research* 40(4): 406–420.

Hsu, T.-H., and M. Lee. 2003. The Refinement of Measuring Consumer Involvement—An Empirical Study. *Competitiveness Review* 13(1): 56–65.

Hu, W., M.M. Veeman, and W.L. Adamowicz. 2005. Labelling Genetically Modified Food: Heterogeneous Consumer Preferences and the Value of Information. *Canadian Journal of Agricultural Economics* 53: 83–102.

Hurst, B. 2005a. Food Packaging Labels—Time to Show and Tell: Management Briefing: Consumer Confidence. *Just-Food* (Jul): 4–13.

Hurst, B. 2005b. The Future of Processed Foods—Management Briefing: Why Do We Rely on Processed Food? *Just-Food* (Apr): 4–6.

Jedidi, K., and Z.J. Zhang. 2002. Augmenting Conjoint Analysis to Estimate Consumer Reservation Price. *Management Science* 48(10): 1350–1368.

Kahneman, D., and A. Tversky. 1979. Prospect Theory: An Analysis of Decision Under Risk. *Econometrica* 46: 171–185.

Kahneman, D., and A. Tversky. 2000. Choices, Values, and Frames. Ch.1 in *Choices, Values, and Frames,* edited by D. Kahneman and A. Tversky, pp. 1–16. New York, New York: Cambridge University Press/Russell Sage Foundation.

Kamstra, M., L. Kramer, and M. Levi. 2003. Winter Blues: A Sad Stock Market Cycle. *American Economic Review* 93: 324–343.

Klein, R.L. 1988. Designating the Right Price for a New Product. *Applied Marketing Research* 28(2): 13–15.

KPMG. 2003. Trends in Retailing 2005—An Outlook for the Food, Fashion and Footwear Sectors. *KPMG Consumer & Industrial Markets.* Cologne, Germany, KPMG Deutsche Treuhand-Gesellschaft AG.

Lähteenmäki, L., and A. Arvola. 2003. Testing Consumer Responses to New Packaging Concepts. Ch. 26 in *Novel Food Packaging Techniques,* edited by R. Ahvenainen, pp. 550–562. Boca Raton, Florida: CRC Press LLC.

Langrehr, F.W., and V.B. Langrehr. 1983. Consumer Acceptance of Item Price Removal: A Survey Study of Milwaukee Shoppers. *Journal of Consumer Affairs* 17(1): 149–171.

Laros, F.J.M., and J.-B.E.M. Steenkamp 2004. Importance of Fear in the Case of Genetically Modified Food. *Psychology & Marketing* 21(11): 889–908.

Lee, B.-K., and W.-N. Lee. 2004. The Effect of Information Overload on Consumer Choice Quality in an On-Line Environment. *Psychology & Marketing* 21(3): 159–183.

Lee, J.K.H., K. Sudhir, and J.H. Steckel. 2002. A Multiple Ideal Point Model: Capturing Multiple Preference Effects from Within an Ideal Point Framework. *Journal of Marketing Research* 39(1): 73–86.

Lewis, H. 2005. Adults' Modern Eating Trends in Major Consuming Countries—2005 Management Briefing: The First Ingredient: Convenient Consumption. *Just-Food* (Jun): 6–9.

Liechty, J.C., D.K.H. Fong, and W.S. DeSarbo. 2005. Dynamic Models Incorporating Individual Heterogeneity: Utility Evolution in Conjoint Analysis. *Marketing Science* 24(2): 285–293.

Light, B. 2004. Kellogg's Goes Online for Consumer Research. *Packaging Digest* 41(7): 40.

Lim, Sonya Seongyeon. 2003. Do Investors Integrate Losses and Segregate Gains? Mental Accounting and Investor Trading Decisions. Ohio State University Working Paper. Ohio State University, Columbus, Ohio.

Löfgren, M., and L. Witell. 2005. Kano's Theory of Attractive Quality and Packaging. *Quality Management Journal* 12(3): 7–20.

Mahon, D., and C. Cowan. 2004. Irish Consumers' Perception of Food Safety Risk in Minced Beef. *British Food Journal* 106(4): 301–312.

Martin, K. 2005. 19th Annual Packaging Trends Survey: Packaging Products Safer, Faster. *Food Engineering* 76(10): 1.

Masten, D.L. 1988. Packaging's Proper Role is to Sell the Product. *Marketing News* 22(2): 16.

McCarthy, M.S., T.B. Heath, and S.J. Milberg. 2001. New Brands Versus Brand Extensions, Attitudes Versus Choice: Experimental Evidence for Theory and Practice. *Marketing Letters* 12(1): 75.

McMath, R.M. 1998a. Don't Get Far Fetched in Your Thinking. *American Demographics* 20(4): 64.

McMath, R.M. 1998b. Image Counts. *American Demographics* 20(5): 64.

Menzies, D. 1998. New, But Maybe Not So Improved: Marketers Should Show Caution When Boasting Product Improvements. *Marketing* 103(34): 10.

Michalek, J.J., F.M.Feinberg, and P.Y. Papalambros. 2005. Linking Marketing and Engineering Product Design Decisions Via Analytical Target Cascading. *Journal of Product Innovation Management* 22(1): 42–62.

Mills, L. 2001. The Fat of the Land. *Marketing Magazine* 106(13): 9.

Mitra, A., M. Hastak, G.T. Ford, and D.J. Ringold. 1999. Can the Educationally Disadvantaged Interpret the FDA-Mandated Nutrition Facts Panel in the Presence of an Implied Health Claim? *Journal of Public Policy & Marketing* 18(1): 106–117.

Mohr, L.A., D. Eroğlu, and P.S. Ellen. 1998. The Development and Testing of a Measure of Skepticism Toward Environmental Claims in Marketers' Communications. *Journal of Consumer Affairs* 32(1): 30–55.

Mukherjee, A., and W.D. Hoyer. 2001. The Effect of Novel Attributes on Product Evaluation. *Journal of Consumer Research* 28(3): 462–472.

Murphy, I.P. 1997. Study: Packaging Important in Trial Purchase. *Marketing News* 31(3): 14.

Nganje, E.W., and S.T.T. Kaitibie. 2005. Multinomial Logit Models Comparing Consumers' and Producers' Risk Perception of Specialty Meat. *Agribusiness* 21(3): 375–390.

Nofsinger, John R. 2005. *The Psychology of Investing*. 2nd ed. Upper Saddle River, New Jersey: Pearson/Prentice-Hall.

Odean, T. 1998. Are Investors Reluctant to Realize their Losses? *Journal of Finance* 53: 1775–1798.

Ollinger, M., and N. Ballenger. 2003. Weighing Incentives for Food Safety in Meat and Poultry. *Amber Waves* (Apr) USDA-ERS.

Oppewal, H., and K. Koelemeijer. 2005. More Choice is Better: Effects of Assortment Size and Composition on Assortment Evaluation. *International Journal of Research in Marketing* 22(1): 45–60.

Paas, L.J., A.A.A. Kuijlen, and T.B.C. Poiesz. 2005. Acquisition Pattern Analysis for Relationship Marketing: A Conceptual and Methodological Redefinition. *Service Industries Journal* 25(5): 661–673.

Peterson, D., and G. Pitz. 1988. Confidence, Uncertainty, and the Use of Information. *Journal of Experimental Psychology* 14: 85–92.

Pieters, R., and L. Warlop. 1999. Visual Attention during Brand Choice: The Impact of Time Pressure and Task Motivation. *International Journal of Research in Marketing* 16(1): 1–16.

Pounder, J. 2005. Make Sure Your Packaging Can Take the R.I.D.E. *Restaurant Hospitality* 89(6): 108–110.

Presson, P., and V. Benassi. 1996. Illusion of Control: A Meta-Analytic Review. *Journal of Social Behavior and Personality* 11: 493–510.

Reast, J.D. 2005. Brand Trust and Brand Extension Acceptance: The Relationship. *Journal of Product & Brand Management* 14(1): 4–13.

Renken, T. 1997. Disaggregate Discrete Choice. *Marketing Research* 9(1): 18–22.

Riquelme, H. 2001. Do Consumers Know What They Want? *Journal of Consumer Marketing* 18(4/5): 437–448.

Samuelson, W., and R. Zeckhauser. 1988. Status Quo Bias in Decision Making. *Journal of Risk and Uncertainty* 1: 7–59.

Schlarbaum, G.G., W.G. Lewellen, and R.C. Lease. 1978. Realized Returns on Common Stock Investments: The Experience of Individual Investors. *Journal of Business* 51: 299–325.

Shapiro, S. 1990. Focus Groups: The First Step in Package Design. *Marketing News* 24(18): 15, 17.

Shefrin, H., and R. Thaler. 1988. The Behavioral Life-Cycle Hypothesis. *Economic Inquiry* 26: 609–643.

Sherlock, M., and T.P. Labuza. 1992. Consumer Perceptions of Consumer Time–Temperature Indicators for Use on Refrigerated Dairy Foods. *Journal of Dairy Science* 75: 3167–3176.

Silayoi, P., and M. Speece. 2004. Packaging and Purchase Decisions: An Exploratory Study on the Impact of Involvement Level and Time Pressure. *British Food Journal* 106(8/9): 607–628.

Silver, D. 2001. Consumer Confidence in Food Safety Takes a Hit. *Restaurants & Institutions* 111(1): 74.

Simonson, Itamar. 2000. The Effect of Purchase Quantity and Timing on Variety-Seeking Behavior. Ch. 41 in *Choices, Values, and Frames*, edited by D. Kahneman and A. Tversky, pp. 735–758. New York, New York: Cambridge University Press/Russell Sage Foundation.

Spaulding, M. 1994. What Consumers Said 10 Surveys Ago. *Packaging* 39(8):44–45.

Spencer, K.C., and D.J. Humphreys. 2003. Argon Packaging Preserves and Enhances Flavor, Freshness, and Shelf Life of Foods. Ch. 20. in *Freshness and Shelf Life of Foods*, ACS Symposium Series 836, edited by K.R. Cadwallader and H. Weenen, pp. 270–291. Washington, DC: American Chemical Society.

Spencer, K.C. 2005. Modified Atmosphere Packaging of Ready-To-Eat Foods. Ch. 12 in *Innovations in Food Packaging*, edited by J. Han, pp. 185–203. San Diego, California: Elsevier Academic Press.

Standard & Poor's. 2005. Foods & Nonalcoholic Beverages. *Standard & Poor's Industry Surveys*. 173(48–1): 1–31.

Stewart, M. 2005. A Fresh, Innovative Look at In-Store Marketing. *Retail World* 58(11): 40.

Thaler, R. 1985. Mental Accounting and Consumer Choice. Marketing Science 4(3): 199–214.

Thaler, Richard H. 2000. Toward a Positive Theory of Consumer Choice. Ch. 15 in *Choices, Values, and Frames*, edited by D. Kahneman and A. Tversky, pp. 269–287. New York, New York: Cambridge University Press/Russell Sage Foundation.

Thompson, S. 2000. Marketers Embrace Latest Health Claims. *Advertising Age* 71(9): 20, 22.

Tingas, P. 2005. A Marriage of Convenience: Smart Marketers Offer Innovative Packaging That Simplifies Consumers' Increasingly Busy Lives. *BrandPackaging* 9(1): 6.

Underwood, R.L., and N.M. Klein. 2002. Packaging as Brand Communication: Effects of Product Pictures on Consumer Responses to the Package and Brand. *Journal of Marketing Theory & Practice* 10(4): 58–68.

Van Heerde, H.J., C.F. Mela, and P. Manchanda. 2004. The Dynamic Effect of Innovation on Market Structure. *Journal of Marketing Research* 41(2): 166–183.

Chapter 9

EUROPEAN STANDPOINT TO ACTIVE PACKAGING—LEGISLATION, AUTHORIZATION, AND COMPLIANCE TESTING

W. D. van Dongen, A. R. de Jong, and M. A. H. Rijk

Introduction

Nonthermally processed foods fit perfectly within the consumer trend of mildly preserved foods, involving an improved level of food taste and quality. On the other hand, milder processed foods are sensitive to microbial or enzymatic food deterioration. Active packaging materials and methods can be applied to preserve the initial high quality of non-thermally processed food. Active packaging is not only restricted to packaging materials used to wrap the food but also refers to any form, shape, or size of active materials and articles.

In USA, Japan, and Australia, active packaging is already being successfully applied to extend the shelf-life food quality and safety. In Europe, the application of active packaging systems is limited. The main reason for this is that until 2004, there were legislative restrictions. As a result of this, there is a lack of knowledge about their acceptability to European consumers and the economic and environmental impact such concepts may have.

This chapter is focused on the current status of European regulations and requirements of the legislative aspects of active packaging.

Since intelligent packaging is in the regulations inextricably related to active packaging, the regulatory aspects of intelligent packaging are also discussed in this chapter. The provisions of the framework directive and the draft regulation on active and intelligent materials are given and discussed. Also attention is given to the requirements that are now being drafted by European Food Safety Authority (EFSA) and which have to be fulfilled for authorization of active and intelligent packaging in Europe.

Finally, recommendations and considerations are given on how to develop and validate dedicated protocols required for compliance testing of active releasing materials.

Active Packaging: Regulations and Requirements for Authorization

Because individual countries in the European Community have not drafted any regulation on the conditions for using active and intelligent packaging, the EU commission has published general requirements and has drafted a specific regulation. The aim is to have safe materials with harmonized requirements all over the European Community. Potential risks of active and intelligent packaging are unacceptable migration, insufficient efficacy, misleading the consumer, or providing incorrect information.

The current and future regulations on active and intelligent packaging have significantly been influenced by an European Commission-funded project known by the acronym of "ACTIPAK" (FAIR-Project CT-98–4170). This project included an inventory of existing active and intelligent packaging and classification of active and intelligent systems in respect of:

- legislation on food contact materials (FCM),
- an evaluation of microbiological safety,
- shelf-life-extending capacity and efficacy of active and intelligent systems, and
- toxicological, economic, and environmental evaluation of active and intelligent systems and recommendations for legislative amendments.

ACTIPAK recommended implementing relevant European regulations and requirements to enable the future application of suitable active and intelligent systems in Europe while maintaining the safety of packed foodstuffs. In addition, it was advised to develop and validate dedicated methods for testing migration behavior of active and intelligent packaging systems.

To be specific, the ACTIPAK project provided recommendations which contributed to the revision of framework Directive 89/109/EEC (Council Directive 89/109/EEC) on FCM. In the new framework Regulation (EC) No. 1935/2004, the use of active and intelligent packaging systems is now included.

A project group under the Nordic Council of Ministers published in 2000 (Fabech *et al.* 2000) a comprehensive report on legislative aspects of active and intelligent food packaging, which also contributed to proposals for new legislation.

A brief overview of the present status of the requirements of active packaging and regulations is given in Table 9.1.

Regulation (EC) No. 1935/2004

In 2004, a new framework Regulation (EC) No. 1935/2004 applicable to all materials intended to come into contact with foods was accepted. In this Regulation, general requirements applicable to all FCM are established. Specific provisions are included to allow the use of active and intelligent materials and articles, and a specific measure on active and intelligent materials was announced.

The definition of active packaging refers to "deliberately" incorporated components with the intention to release or absorb substances. This distinguishes active materials from passive packaging materials, which in a few cases may have an effect on the food, but to which compounds are added for other reasons, for example, as a monomer or an additive. The definition also excludes all packaging materials from natural sources. For instance, wooden barrels are therefore not subject to the provisions on active materials. The substances released from these are not deliberately added to food. But if a wood extract were to be incorporated into an active packaging system, then this would fall under the general and specific measures on active materials and articles. Intelligent FCM and articles are materials and articles which monitor the condition of packaged food or the environment surrounding the food.

Article 3 of the framework regulation requires that FCM shall not transfer their constituents to foods in quantities which could "endanger human health, or bring about an unacceptable change in the composition of the food, or bring about a deterioration in the organoleptic characteristics."

Table 9.1. Overview of current regulations and requirements relevant for active and intelligent packaging.

Regulation	Regulations and Requirements Issues Related to Active and Intelligent Packaging
Framework Regulation (EC) No. 1935/2004 Abstract State that all FCM may: • not endanger human health • no unacceptable change in composition • no deterioration of the organoleptic characteristics • not mislead the consumer	General • Authorization of A/I materials subject to EFSA evaluation. These are in accordance with "general food law" (Regulation (EC) No. 178/2002) • Compliance must be demonstrated to relevant authorities • Active materials change the composition or organoleptic properties of food. To account for this, specific provisions are included in Article 4 Article 4 • *Allows changes* in food composition/ organoleptic properties, provided they comply with 89/107/EEC on food additives • *Misleading* consumer (intelligent) and masking of spoilage not allowed • *Labeling* of precense A/I materials/ released substances/nonedible parts
Draft regulation on A/I materials (EMB/973 Rev. 5B) Abstract Final regulation most likely will not differ significantly from Regulation (EC) No. 1935/2004. Deals with authorization of A/I components and can be considered as a starting point on requirements on active and intelligent packaging	Authorization • Draft regulation requires individual authorization of A/I components • *Listing.* A/I components are inserted on a list of authorized components • *Suitability.* Demands that active and intelligent packaging "are suitable and effective for the intended purpose" and are authorized • *Certification.* A/I materials must demonstrate to be safe as FCM under specified conditions. Documentation that proofs the validity of the certificated must be available

Table 9.1.   (*Continued*)

Regulation	Regulations and Requirements Issues Related to Active and Intelligent Packaging
	Migration • *Carriers.* Should comply with safety requirements of framework Regulation (EC) No. 1935/2004 and implemented EU/national measures • *Active releasing.* Substances should *not* be included in overall migration, special protocols are needed. Specific migration limits may be exceeded, provided final food complies with processed foods rules and restrictions
Requirements for authorization—EFSA guidelines Abstract General guidelines for food contact materials (EFSA Note for Guidance, 2001), in particular for plastics. Guidelines not generally applicable to A/I EFSA guidance for A/I under preparation is available at the moment the regulation on active and intelligent packaging is ready for implementation	Opinions of EFSA are based on a risk assessment. EFSA will require all data needed to make a proper safety assessment. In Figure 9.1, a flow scheme related to the authorization procedure is depicted • *Releasing materials.* Authorization as a food additive is required. Released component must show effectiveness to comply with food additive requirements. Carrier of releasing material should notmigrate at unacceptable quantities • *Absorbing materials.* Focus will be on the toxicological properties and quantities of migrants. Efficiency/ capacity of absorber will be considered

(continued)

Table 9.1.   Overview of current regulations and requirements relevant for active and intelligent packaging. (*Continued*)

Regulation	Regulations and Requirements Issues Related to Active and Intelligent Packaging
	• *Intelligent materials.* Must demonstrate reliability of provided information. Unauthorized substances may be used behind a barrier, provided the substances are not MCR according to Directive 67/548/EEC. In case of food contact: levels and toxic properties of migrants must be evaluated
Relevant measures to be considered	• *General food law (Regulation No.178/2002).* Sets general requirements in respect of food and feed safety • *Food additives (89/107 EEC).* Releasing systems are subject to this directive • *Biocides (98/8/EC)* note that biocides are not food additives • *Labeling (2000/13/EC).* Released substances should be labeled • *Hygiene (Reg. no.852/2004).* Active materials to assure food integrity • *General product safety (2001/95/EC)* • *Weight and volume control (1976/211/EEC)*

In particular, releasing materials cannot meet these requirements as they are designed to change the composition or the organoleptic properties of the food. Absorbing materials may also change the composition or organoleptic properties of the food. Therefore, a special article 4 has been inserted which allows changes in the composition or organoleptic characteristics of the food, provided the changes comply with the provisions of Directive 89/107/EEC (see reference) on food additives and its related implementing measures. In the absence of Community measures, national provisions shall be applicable. Inserting this provision took away the hurdle, in the old framework Directive 89/109/EEC, to the introduction of active packaging with a releasing function. Additional requirements on active and intelligent materials are related to misleading the consumer and to labeling. Active packaging used to change the composition of food or its organoleptic properties in order to mask spoilage of that food is not acceptable. In many cases, there will not be a clear line between misleading and improving the organoleptic properties of a packaged food. For instance, is the removal of aldehyde components from nuts misleading or improving?

The regulation introduces the EFSA and its role, including procedures and time frames. These are in accordance with "general food law" (Regulation (EC) No. 178/2002). EFSA has to be consulted on issues affecting public health. This means also that the authorization of active and intelligent materials will be subject to an EFSA evaluation.

If required in a specific measure, relevant FCM shall be accompanied with a declaration of compliance, while appropriate documents shall be provided to relevant authorities to demonstrate such compliance. As most of the requirements in the framework regulation are applicable to all FCM, active and intelligent materials are subject to these rules as they can be considered FCM. In some cases, there may be no direct contact with the food, for example, intelligent packaging positioned on the outside of the primary package, but they are subject to the framework regulation for the reliability of the information provided to the consumer.

Draft Regulation on Active and Intelligent Materials and Articles

A specific regulation (EMB/973 Rev. 5B) on active and intelligent packaging is approaching its final stage. The final regulation most likely will not differ significantly from the document now available. Therefore, the document can be taken as a starting point on requirements on active and intelligent packaging.

The draft regulation in particular deals with the authorization procedure for active and intelligent "components." It should be taken into account that active and intelligent packaging consists of two parts. First the active or intelligent components, and the other part is the so-called carrier or passive part containing the component. In the example of an ethanol releaser, the ethanol is absorbed on a silica gel which in turn is packaged in a paper or plastic sachet. The ethanol is defined as the active component, which is subject to authorization. The silica gel and the sachet form the passive part and should comply with safety requirements as defined in the framework regulation and implemented EU or national measures.

Migration of FCM is subject to EU or national regulations. Overall migration and specific migration limits are established in the various regulations. These limits are set to assure inertness and safety of the FCM. Active releasing packaging is not designed to be inert and will in many cases exceed the overall migration limit and in some cases the specific migration limits set for FCM. Specific migration limits may be exceeded, provided the final food complies with the rules and restrictions applicable to processed foods. Therefore, a substance released on purpose from an active packaging material should not be included in overall migration, making CEN methods EN 1186, Parts 1–15 (EN 1186, Parts 1–15) unsuitable. Thus, for active packaging, dedicated protocols may be needed for the determination of the overall migration of the passive part only.

Finally, the draft regulation requires that active and intelligent packaging "are suitable and effective for the intended purpose" and that the active and intelligent components are authorized.

Authorization will be granted after a positive opinion of EFSA and will only be valid to the applicant for an authorization. The authorization will be valid for a period of ten years and may be renewed for another period of ten years. The authorization will be published in a Decision to the applicant. In addition, the active and intelligent components will be inserted in a list of authorized components.

In line with Regulation (EC) No. 1935/2004, a requirement concerning a declaration of compliance and the availability of appropriate documentation has been confirmed in the draft regulation. It means that for any active and intelligent material, a statement shall be provided that certifies that the material is safe to be used in contact with food under specified conditions of contact. To support such a statement, the certifier shall have documentation that can prove the validity of the certificated. These documents shall be available to relevant authorities for inspections. In many cases, this will include analytical data on, for example, migration, total release, and effectivity of the active and intelligent of active components.

Requirements for Authorization—EFSA Guidelines

EFSA has been appointed to advise the European Commission on the safety of substances to come into contact with foods. Opinions of

EFSA are based on a risk assessment. In general, the conclusions given in an opinion will be adopted by the Commission, although occasionally the Commission may decide to deviate from the EFSA opinion as a risk management issue. In Figure 9.1, a flow scheme related to the authorization procedure is depicted.

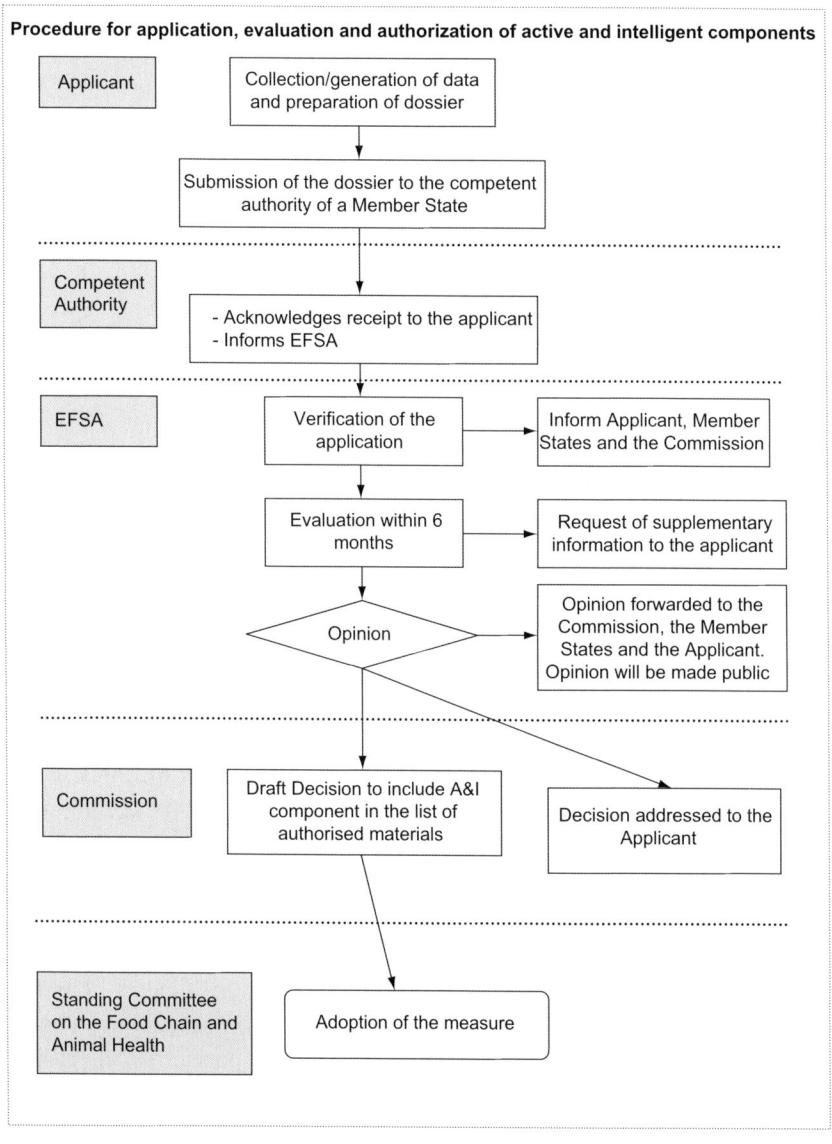

Figure 9.1.   Procedure for application, evaluation, and authorization of active and intelligent components.

The Scientific Committee of Food (SCF), the predecessor of the EFSA, established guidelines for FCM and in particular for plastics. EFSA's working group on FCM provided detailed explanatory guidance (AFC-FCM-WG explanatory guidance, 2005) in the Note for Guidance. The guidelines are not generally applicable to active and intelligent packaging, and EFSA will publish additional guidelines and explanatory guidance that may support applicants when drafting an application and developing a testing protocol. This guidance should be available at the moment the regulation on active and intelligent packaging is ready for implementation.

Only a rough and predictive overview of the guidelines is feasible at this stage. EFSA will require all data needed to make a proper safety assessment, and despite any guidelines, they will always be authorized to ask for additional information.

It is likely that an application should contain:

- General information on the identity of the applicant.
- Summary document which summarizes all the information provided.
- Technical dossier containing a description of the active or intelligent material, its function, and the active or intelligent components. Additional information may be required, but this could depend on the type of active or intelligent packaging.

For releasing materials, there will likely be a focus on the releasing component and its authorization as a food additive, including any quantitative restriction or a restriction on the types of food. The information on efficacy may be important, for example, in the case of a released preservative, the final efficacy in the food should be demonstrated. A general rule may be considered that if the released component shows insufficient or no technical effect on the food, then the food additive does not comply with the requirements on food additives (Council Directives 89/107/EEC; 94/34/EC; Regulation (EC) No. 1882/2003) or any other relevant regulation on the composition of food and its additives, for example, the requirements on food flavors (Council Directives 88/388/EEC and 91/71/EEC). As a consequence, such a material may not obtain a favorable opinion. The carrier of the releasing substance will not be part of an authorization. The safety of the carrier is the responsibility of the producer and the final user. In many cases, the carrier may be subject to other provisions on FCM.

In principle, the carrier should be inert and should not migrate to the food at an unacceptable concentration.

For absorbing materials, the focus will be on the toxicological properties and quantities of components that (unintentionally) migrate into the food. Depending on the type of material used for the absorber, the relevant directive will be applied, for example, if the absorber consists of plastic only, then the plastics Directive 2002/72/EC (Corrigendum to Commission Directive 2002/72/EC) will be valid. Many active materials are composed of various types of materials, for example, plastic, paper, metal, and adhesive. No harmonized EU regulation exists on these materials, and therefore they are subject to the framework regulation (Regulation (EC) No. 1935/2004) and relevant national provisions of EU Member States. The efficiency or capacity of the absorber will almost certainly be considered in the final evaluation. As an example, the use of an oxygen absorber which has insufficient capacity to decrease and maintain the oxygen content at a low concentration may at least mislead the consumer or create in the worst case an even more dangerous situation due to the overgrowth of specific microorganisms. Also, the possible growth of anaerobic bacteria present on some foodstuffs may be a reason not to apply an oxygen scavenger. It could be questioned whether this can be covered in an authorization. It may be better to maintain the responsibility for the final food safety with the food packers who are familiar with their food products and possible problems.

Intelligent materials will probably need a demonstration of the reliability of the information provided, to comply with the respective requirement in the draft regulation (EMB/973 Rev. 5B). In respect of unintended migration, a distinction may be feasible for materials that are positioned inside or outside the primary packaging. In the latter case, the actual migration of any intelligent component may be negligible, and it would be logical to adapt the need for toxicological information to the "no migration" or "functional barrier" principle. This would mean that unauthorized substances can be used behind a barrier layer, provided the substances are not mutagenic, carcinogenic, or toxic to reproduction according to Council Directive 67/548/EEC. If the intelligent material may come into contact with the food, then the intelligent ingredients will need an evaluation based on migration level and toxicological properties as outlined in the Note for Guidance (EFSA Note for Guidance, 2001; AFC-FCM-WG, 2005).

General Safety

Many EU directives and regulations related to food and food safety have been published. All relevant regulations have been extensively discussed before (De Kruijf and Rijk). Without being exhaustive, the following European or Member States regulations are taken into account in an evaluation procedure performed by EFSA:

- FCM
- food additives
- flavorings
- hygiene
- biocides
- labeling
- general product safety
- misleading claims
- weight and volume

As a general rule, all the regulations related to food have the requirement that the final food should be safe for consumption. This may relate to chemical composition or to microbiological deterioration. Chemical composition may refer to food additives (Council Directive 89/107/EEC), flavoring substances (Council Directives 88/388/EEC and 91/71/EEC), or sweeteners (Council Directive 94/35/EC). If these substances are added via active packaging, then the final food must comply with the relevant requirements for food production. The regulation on hygiene (Council Directive 93/43/EEC) requires that all measures shall be taken to assure the wholesomeness of the food. Active packaging may be a useful tool to comply with this requirement as it is usually applied to extend the shelf life of the food or at least to maintain the quality of food during its shelf life. The regulation on biocides (Council Directive 98/8/EC) is actually not applicable as biocides are not allowed in processed food. Only substances which are authorized as preservatives in food may be applied. FCM with antimicrobial surfaces (e.g., after incorporation of silver ions in FCM) are frequently, but incorrectly, referred to as active packaging. The antimicrobial surface has no effect on the food itself, and therefore it is excluded from the definition of active packaging.

Products shall be safe as established in Directive 2001/95/EC (Council Directives 2001/95/EC). "Safe products" means that under normal or reasonably foreseeable conditions of use, the product does not present any risk or only the minimal risks compatible with the product's use. Some active packaging is designed to absorb components from the packaged food, for example, a moisture absorber. When using this type of absorber, the final weight of the food available for consumption may change. Council Directives 75/106/EEC and 76/211/EEC include labeling of weight and tolerances. When part of the food is absorbed by an absorber, then the final weight may no longer comply with the intended and declared food weight.

Besides the framework Regulation 1935/2004 and the specific Directive 2002/72/EC (Corrigendum to Commission Directive 2002/72/EC) on FCM, other directives on FCM have been adopted and may be relevant in assuring the safety of active or intelligent packaging. This concerns directives on regenerated plastics (Council Directives 93/10/EEC, 93/111/EC, and 2004/14/EC), ceramics (Council Directive 84/500/EC), and a regulation on certain epoxy compounds (Commission Regulation (EC) No. 1895/2005). In the absence of harmonized EU regulations, existing national regulations may be applicable and may be useful to demonstrate safety of active or intelligent packaging.

Labeling

In judging safety aspects, the characteristics of the product, presentation, labeling instructions, and the category of consumers, especially children, should be considered. Also the shape and size of active or intelligent packaging in the form of, for example, a sachet that is packed together with the food may give reasons for concern. Controls must stop possible dangers to human health to prevent consumers trying to eat sachets filled with active compounds. Also a visually attractive objects packed with the food can be mixed up with foodstuffs. Appropriate labeling will be essential. Labeling of foodstuffs is meant to give consumers information on the composition of the food and hence protect their interests.

Labeling appears to be a complex issue as there are many EU directives and regulations that include requirements on labeling. Food labeling may concern the food, food additives or FCM, as well as active and intelligent packaging. For active packaging, information on the permitted use and the identity and quantity of the released substance have to be provided in order to allow a food packer to comply with any restriction. Materials that may be mistaken as a part of the food such as loose sachets must be labeled using the symbol for a nonedible part (Figure 9.2).

DO NOT EAT
NE PAS MANGER

Figure 9.2.   Symbol for a nonedible part of the food.

Commission Directive 2000/13/EC deals with labeling, presentation, and advertising of foodstuffs and is applicable to all foodstuffs intended for sale to consumers or caterers. Also Directive 89/107/EEC sets requirements on labeling of food additives. In principle, all substances used in the manufacture or preparation of foodstuffs and still present in the finished product should be declared on the label, in order to inform the consumer about the substances present. It is of no interest how or when the substances are added to the food, and therefore substances released from an active packaging system should be declared.

Modified atmosphere gases are regulated as food additives (Directive 95/2/EEC). Gases that may be added are listed. When modified gas packaging is applied, this should be labeled. However, there are no provisions for the removal of gases from the packed food. This means that a CO_2 releasing system falls within the definition of a food additive. But an oxygen scavenger does not comply with the definition as the oxygen is removed. The final gas composition in a modified atmosphere may be comparable to that in packaged food using an oxygen absorber, but it is not yet clear whether there is any regulation that would prevent the use of oxygen absorbers. On the other hand, there is no prohibition on the removal of the oxygen from packaged food and no labeling requirement for the modified atmosphere. Only Regulation 1935/2004 requires that the use or presence of an active device, extending the shelf life of the food, should be labeled.

Migration from Active and Intelligent Packaging into Foodstuffs

Basic Rules

Active and intelligent packaging are considered to be FCM. This means that the migration of packaging substances should be examined according to the existing EU directives or national provisions. At the EU level, the framework Regulation 1935/2004 sets the general requirements of which article 3, stating that the packaging material should not endanger human health, is the most important. Following up the framework regulation, specific directives on plastics (2002/72/EC), regenerated cellulose (93/10/EEC), and ceramics (84/500/EEC) have been published and implemented in national legislation in the EU. The plastics directive contains a positive list of monomers and an incomplete list of additives for the use in plastics. To support testing of plastic materials for compliance with the plastic directive, two additional directives are available. Directive 82/711/EEC as last amended by Directive 97/48/EC sets requirements (time–temperature conditions, selection of simulants) for testing of plastic materials and articles with food simulants. Directive 85/572/EEC indicates the simulants that shall be used for specified foods or groups of foods. Regulation of chemical migration from plastics into food and drink has developed over many years. EU regulatory control of active and intelligent packaging will therefore have to fit in with this well-established system of controls.

To support the analyst in applying such controls, CEN (the European Standardization Commission) has in TC 194 adopted and validated analytical methods for the determination of the overall migration (EN 1186, Parts 1–15) and the migration of some specific substances. These methods are intended to be applied for testing plastic materials and articles. At the national level, for example, in The Netherlands (Verpakkingen-en gebruiksartikelenbesluit van 30 mei, 2005), the methods and simulants may also be used to demonstrate compliance with national regulation of nonplastic or multilayer materials composed of plastics and nonplastics (e.g., plastic on paper and coating on metal).

The principle of the methods is rather simple, but analytical problems are frequently encountered. An FCM is brought into contact with the selected simulant(s) under selected conditions of time and temperature. Homogeneous materials are contacted by submersion, while multilayers or thin films are often single-sided contacted with the food simulant. After the contact period, the simulant is separated from the FCM, and the overall migration is determined gravimetrically, while specific migration is determined using a suitable analytical method like gas or liquid chromatography with spectroscopic detection. The determination of the overall migration in olive oil is more complex and sensitive to analytical and systematic errors. An indirect method is required to determine overall migration (EN 1186 Parts 1–15).

Migration of Active and Intelligent Packaging

Depending on its function, active and intelligent packaging have a varying shape, size, and composition. Active and intelligent materials can be applied in several manners in packaging of food:

- active or intelligent packaging which is also the primary packaging;
- active packaging as a releasing material;
- active material as an absorber;
- intelligent material inside the primary packaging; or
- intelligent material outside the primary packaging.

Another variable to consider is the type of food in contact with the active or intelligent packaging. One can distinguish food contact with a liquid (e.g., for an oxygen-scavenging crown cork for beer), dry contact (e.g., a preservative releaser for buns), or with a semi-solid (e.g., an oxygen scavenger for processed ham).

In many cases, conventional migration tests, that is, single-sided contact, cannot be applied due to the size or application of the active material. Within the ACTIPAK project (FAIR-Project CT-98-4170), migration experiments were performed by total immersion of active and intelligent packaging into various simulants. It appeared that in many cases overall migration was exceeded (Rijk and De Kruijf, 2003), and it was concluded that determination of overall migration from active and intelligent packaging using conventional methods is not applicable in many cases. There is a need for dedicated tests that simulate better the conditions of contact for some types of active and intelligent packaging (Lopez-Cervantes *et al.*, 2003) to allow proper judgment of its suitability. For each type of application, almost a complete different approach of migration testing is required.

Dedicated Migration Testing of Active and Intelligent Packaging

Typical examples of active and intelligent packaging functions in primary packaging are an oxygen scavenger with the active ingredients incorporated in the polymer backbone or a casing containing a smoke flavor to be released to the packaged food. Certainly in new developments, the primary packaging will more often be combined with the active or intelligent function. The composition of the packaging should comply with EU regulations, relevant national regulations, and with the active or intelligent component mentioned in the authorization. Compliance of overall and specific migration with EU limits can be examined according to Council Directives 82/711/EEC and 85/572/EEC applying the conventional test methods outlined in the CEN methods EN 1186 (Parts 1–15) and EN 13130 (Parts 1–28). Both single-sided contact and total immersion may be applied. For verifying compliance with the restrictions, the conventional assumption that 6 dm² of packaging is in contact with 1 kg of food will generally be applicable, as it is for conventional packaging materials. Available migration test methods may not be suitable for the determination of migration from releasing materials. For overall migration, the release of the active component may be much greater than the overall migration limit. According to the draft regulation on active and intelligent packaging (EMB/973 Rev. 5B), the release of the active component should be excluded from overall migration. From an analytical point of view, there are two options. One could determine the overall migration from the packaging material which does not contain the releasing substance as it is done for conventional passive packaging. Alternatively, overall migration of the active packaging can be determined, and from this the migration of the active component can be deducted. In this way, migration of the passive part of the packaging can be determined. Of course, this will introduce a relatively large analytical error depending on the amount and nature of the released component. This method may only be applicable for a few types of active releasing packaging.

Active Packaging as a Releasing Material

Active packaging is often included in the content of the primary packaging as a small sachet, box, or label and cannot be compared to conventional (passive) packaging materials. For conventional packaging materials, the usual ratio of 1 kg of food in contact with 6 dm² is applied. For sachets and the like, the surface to volume ratio is usually much smaller. Active releasing packaging can be composed of very different materials. An ethanol releaser consists of ethanol absorbed onto a powdered silica gel carrier in a plastic-coated paper sachet. This type of active packaging is not intended for contact with liquid foodstuffs, but for contact with dry and semi-solid foods. For this type of packaging, it should be realized that article 3 of the framework Regulation (EC) No. 1935/2004 applies as well as relevant provisions on food additives. It is difficult—if possible at all—to determine the migration of the carrier using model tests with simulants. On the other hand, there must not be unacceptable migration from the carrier to the food. The most simple and efficient way to certify compliance with the framework regulation on FCM is to use approved materials in the manufacture of active packaging. This means composition and migration (overall and specific) should comply with relevant EU regulations or national provision. This, however, is not feasible for enforcement purposes. Enforcement authorities could determine specific migration of the carrier constituents in the actual food. In simulating tests, the use of the "dry simulant" MPPO may be useful for the determination of specific migration of substances. For semi-solid contact, the so-called sandwich method, which is still under development and validation (TNO, Zeist, Netherlands) may be applied, provided conditions of contact are selected with great care.

*Active Material as an Absorbent or Intelligent
Material Inside Primary Packaging*
Active absorbers and intelligent packaging by definition have no intentional migration. Nevertheless, migration may occur as this is noticed from passive packaging. This type of active and intelligent packaging shall comply, as a minimum, with the framework regulation on FCM. Also the migration of active or intelligent components has to comply with this framework regulation. For testing of compliance, in the first instance, the composition of the active and intelligent component should be covered by an authorization, and second the composition of the packaging of the active or intelligent component should comply with the relevant EU or national regulations on FCM. Chemicals used as a carrier of the active or intelligent component may not be subject to a specific regulation, for example, presence of sodium chloride in an iron-based oxygen scavenger (discussed below). Following the principle of inertness of FCM, the carrier should not migrate to the food.

Rules for inertness and migration behavior of FCM are well defined. But how to demonstrate this is not yet established and will be complex, because of the almost endless variations of types of food, shape, and size of active packaging and conditions of contact. Many types of active and intelligent packaging inserted in the primary packaging together with the food cannot be tested by conventional means, and dedicated test protocols are needed. Up till now, only a few protocols have been found suitable for dedicated testing of some active and intelligent packaging in contact with semi-solid foods.

Testing of Iron Oxide-Based Oxygen Absorbers
for Overall and Specific Migration
Oxygen absorbers used for shelf-life extension of semi-solid foods, like cooked ham, should be tested with 3% acetic acid when the pH of the ham is less than 4.5. Total immersion of the oxygen scavenger is not realistic. Performing a submersion test results in a simulant colored brown with iron oxide. As this brown coloring does not occur on the food, the contact mode by total submersion is obviously different from contact in real life. In Directive 97/48/EC, provisions are made to deviate from the prescribed test conditions (time and temperature) when physical or other changes are observed that do not occur in real-life contact with the food. This opens the way to using dedicated methods. Experiments with satisfying results have been obtained by using a so-called sandwich method as depicted in Figure 9.3.

In this experiment, a sample of active packaging (label or sachet) was placed on a glass plate and covered with a stack of filter paper that first had been extracted extensively with the test simulant. Free-flowing simulant should be removed by allowing the simulant to drip for about 30 seconds. Then the filter paper is covered with another glass plate and, in case the weight of the food should be simulated, a weight can be put on top of the upper glass plate. To avoid evaporation of the stimulant, the whole package can be wrapped with a plastic film or aluminum foil. The packaged is stored under selected conditions of time and temperature. After the storage period, the filter paper is analyzed for overall or specific migration. For the overall migration, the filter paper is extracted in a soxhlet apparatus using the test simulant or a more suitable solvent. The extraction residue is determined gravimetrically. Correction of the blank value for the filter paper is necessary. For the determination of the specific migration of substances, the soxhlet extract can be analyzed. Volatile substances

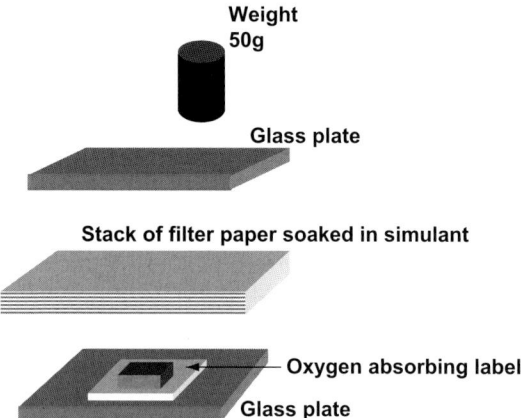

Figure 9.3.   Council Directive (and European Parliament) 95/2/EC Exploded view of dedicated testing of active packaging labels or sachets.

can be analyzed using filter paper without extraction. For non volatile substances that give problematic recovery (in this case of iron ions), the filter paper may be combusted, and the migration of iron can be determined using atomic absorption spectrometry. This procedure is applicable to various types of active absorbing and intelligent packaging which have a more or less flat shape like a sachet, label, or pad.

A moisture-absorbing pad may be tested with this protocol. In principle, such a pad will absorb the meat drip into its fibers or in a cross-linked polymer. Due to the absorption capacity, the migration of the absorber's constituents is not likely. However, in case the absorber starts to become saturated, migration is more likely. To simulate this, an absorber should first be saturated with water to approximately 80% of its capacity before migration testing.

Active and intelligent packaging in contact with liquid foods can be tested by submersion in simulant as this is equivalent to real-use conditions. Packaging used in dry foods should not be tested with liquid food simulants. If testing is considered necessary, then the use of MPPO may be suitable. MPPO is dry, has a large surface area, and has a strong absorptive capacity. Any migration into dry foods occurring in real life can be detected also by using MPPO, which may be analyzed after extraction or directly by thermodesorption.

Intelligent Material is Outside the Primary Packaging
Some intelligent packaging [e.g., time–temperature indicators (TTI)] is placed *in situ* on the outside of the packaging. Taking into account the

construction of most TTIs and the barrier of the primary food packaging, it is assumed that a functional barrier prevents the migration of any of the intelligent components. Migration testing should then not be necessary. Of course, the composition of the authorized component and the carrier shall be in compliance with regulatory requirements.

Expression of Migration Results
Many forms of active and intelligent packaging, not used as primary packaging, are designed to have little or no contact with food. For instance, an oxygen absorber, based on iron oxide, applied in the shape of a sachet has only a contact area of a couple of square centimeters. The conventional surface to volume ratio of $6\,dm^2/kg$ food will never be achieved. Similar arguments are valid for many types of intelligent packaging. Plastics Directive 2002/72/EC allows the use of the real ratio of surface to volume. Therefore, migration should be expressed in mg/subject and divided by the quantity (kg) of food in contact with the subject. This should be done before using overall or specific migration limits listed in relevant regulations. It should be borne in mind that there may be a contribution to overall or specific migration originating from the primary packaging.

Conclusions

The new regulation will authorize the use of active packaging, provided the packaging can be shown to enhance the safety, quality, and shelf life of the packaged foods. The new framework regulation (EC) No. 1935/2004 directive allows the application of active packaging systems. The main hurdle to the use of active and intelligent packaging in the EU, that is, migration of active components into food, has been taken away by the framework regulation by allowing changes in the composition or organoleptic characteristics of the food. If they are food additives, then they may change the composition of the food, provided the final food complies with the food additive regulation. The regulation will authorize the use of active packaging, provided the packaging can be shown to enhance the safety, quality, and shelf life of the packaged foods.

A specific regulation for A&I packaging is approaching its final stage. The final regulation most likely will not differ significantly from the document now available. Other relevant regulations, for example, regulations for food additives, flavoring, biocides, pesticides, food hygiene, labeling, product safety, and waste, generally do not form a serious hurdle to the introduction of active and intelligent packaging systems. The directive on food hygiene is even an incentive to the use of active and intelligent packaging systems. The aim to have safe materials with harmonized requirements all over the European Community has been secured by the new framework regulation.

An outcome of the ACTIPAK project did demonstrate that acceptance is not gained purely on the extension of shelf life, but also the nature of the applied technology and releasing chemicals are very essential for consumer acceptance. Releasing compounds which are for instance of nonbiological origin or have gained in other applications a negative image are less acceptable for the consumer then for instance the use of vitamin C as releasing antioxidant.

Despite the consumer acceptance, hurdles that have to be overcome in the near future, there is a strong view that active packaging will become a technical tool in the European market with a high potential, covering both a shelf-life-extending capacity for nonthermally processed food products and a transparent communication to consumers.

References

AFC-FCM-WG explanatory guidance of the SCF guidelines for food contact materials. Note for Guidance, Chapter III, Updated on 28/9/2005. http://www.efsa.eu.int/science/afc/afc_guidance/722/noteforguidance1.pdf.

Commission Directive 91/71/EEC of 16 January 1991 completing Council Directive 88/388/EEC on the approximation of the laws of the Member States relating to flavourings for use in foodstuffs and to source materials for their production, Official Journal, L 042, 15/02/1991, 0025–0026.

Commission Directive 93/10/EEC of 15 March 1993 relating to materials and articles made of regenerated cellulose film intended to come into contact with foodstuffs. Official Journal, L93, 17/04/1993, 0027–0036.

Commission Directive 93/111/EC of 10 December 1993, amending Directive 93/10/EEC relating to materials and articles made of regenerated cellulose film intended to come into contact with foodstuffs. Official Journal, L310, 14/12/1993, 0041.

Commission Directive 97/48/EC of 29 July 1997 amending for the second time Council Directive 82/711/EEC laying down the basic rules necessary for testing migration of constituents of plastic materials and articles intended to come into contact with foodstuffs. Official Journal, 222 of 12/8/1997, 10–15.

Commission Directive 2004/14/EC of 29 January 2004, amending Directive 93/10/EEC relating to materials and articles made of regenerated cellulose film intended to come into contact with foodstuffs. Official Journal, L027, 30/1/2004, 48–51.

Commission Regulation (EC) No. 1895/2005 of 18 November 2005 on the restriction of use of certain epoxy derivatives in materials and articles intended to come into contact with food, Official Journal, L 302, 28–32.

Corrigendum to Commission Directive 2002/72/EC of 6 August 2002 relating to plastic materials and articles intended to come into contact with foodstuffs (Official Journal, L220, 15/8/2002, 0018–0058.), Official Journal, L 0392, 13/02/2003, 01–42.

Council Directive 67/548/EEC of 27 June 1967 on the approximation of laws, regulations and administrative provisions relating to the classification, packaging and labelling of dangerous substances. Official Journal 196, 16/08/1967, P. 0001–0098.

Council Directive 75/106/EEC of 19 December 1974 on the approximation of the laws of the Member States relating to the making-up by volume of certain prepackaged liquids. Official Journal, L 042, 15/02/1975, 0001–0013.

Council Directive 76/211/EEC of 20 January 1976 on the approximation of the laws of the Member States relating to the making-up by weight or by volume of certain prepackaged products. Official Journal, L 046, 21/02/1976, 0001–0011.

Council Directive 82/711/EEC of 18 October 1982 laying down the basic rules necessary for testing migration of constituents of plastic materials and articles intended to come into contact with foodstuffs. Official Journal, 297 of 23/10/1982, 26–30.

Council Directive 84/500/EC of 15 October 1984 on the approximation of the laws of the Member States relating to ceramic articles intended to come into contact with foodstuffs. Official Journal, L277, 20/10/1984, 0012–0016.

Council Directive 85/572/EEC of 18 December 1985 laying down the list of simulants to be used for testing of constituents of plastic materials and articles intended to come into contact with foodstuffs. Official Journal, 372 of 31/12/1985, 14–21.

Council Directive 88/388/EEC of 22 June 1988 on the approximation of the laws of the member states relating to flavourings for use in foodstuffs and to source materials for their production. Official Journal, L 184, 15/07/1988, 0061–0066.

Council Directive 89/107/EEC of 21 December 1988 on the approximation of the laws of the Member States concerning food additives authorised for use in foodstuffs intended for human consumption. Official Journal, L 040, 11/02/1989, 0027–0033.

Council Directive 89/109/EEC of 21 December 1988 on the approximation of the laws of the Member States relating to materials and articles intended to come into contact with foodstuffs. Official Journal, L 040, 11/02/1989, 0038–0044.

Council Directive (and European Parliament) 94/35/EC of 30 June 1994 on sweeteners for use in foodstuffs. Official Journal 237, 10/9/1994, 3. Last updated by Directive 2003/115/EC of the European Parliament and of the Council of 22 December 2003. Official Journal, 24, p. 65.

Council Directive 93/43/EEC of 14 June 1993 on the hygiene of foodstuffs. Official Journal, L 175, 19/07/1993, 0001–0011.

Council Directive (and European Parliament) 94/34/EC of 30 June 1994 amending Directive 89/107/EEC on the approximation of the laws of Member States concerning food additives authorised for use in foodstuffs intended for human consumption, Official Journal, L 237, 10/09/1994, 0001–0002.

Council Directive (and European Parliament) 95/2/EC of 20 February 1995 on additives other than colours and sweeteners. Official Journal, 61, 18/3/1995, p. 1. Last amended by Directive 98/72/EC of the European Parliament and of the Council of 15 October 1998. Official Journal, 295 of 4/11/1998, p. 18.

Council Directive (and European Parliament) 98/8/EC of 16 February 1998 on the placing of biocidal products on the market. Official Journal, L 123, 24/04/1998, 0001–0063.

Council Directive 2000/13/EC of (and European Parliament) of 20 March 2000 on the approximation of the laws of the Member States relating to the labelling, presentation and advertising of foodstuffs. Official Journal, L109, 6/5/2000, 29–42.

Council Directive 2001/95/EC of 3 December 2001 on general product safety (text with EEA relevance). Official Journal, L 11, 15/01/2002, 0004–0017.

De Kruijf, N. and Rijk, R. 'Legislative issues relating to active and intelligent packaging', Novel Food Packaging Techniques, Woodhead Publishing Limited ISBN 1 85573 675 6.[P15]

EFSA Note for Guidance, Chapter II, Updated on 13 December 2001, Guidelines of the Scientific Committee on Food for an application for safety assessment of a substance to be used in food contact materials prior to its authorisation. http://www.efsa.eu.int/science/afc/afc_guidance/722/noteforguidance1.pdf.

EMB/973 Rev. 5B, Working Document on a Draft Regulation on active and intelligent materials and articles intended to come into contact with foodstuffs. Version updated to 16 May 2005.

EN 1186, Parts 1–15, Materials and articles in contact with foodstuffs—Plastics.

EN 13130, Parts 1–28, Materials and articles in contact with foodstuffs—Plastics substances subject to limitation.

Fabech, B., Hellstrøm, T., Henrysdotter, G., Hjulmand-Lassen, M., Nilsson, J., Rüdinger, L., Sipiläinen-Malm, T., Solli, E., Svensson, K., Thorkelsson, À.E., and Tuomaala, V., Active and intelligent food packaging—A Nordic report on legislative aspects, Copenhagen, Nordic Council of Ministers, 2000.

FAIR-Project PL 98-4170. 'Actipak': Evaluating safety, effectiveness, economic-environmental impact and consumer acceptance of active and intelligent packaging. Duration 1998–2001. Final report: 2003.

Lopez-Cervantes, J., Sanchez-Machado, D.L., Pastorelli, Rijk, R., S., and Paseiro Losada, P., Evaluating the migration of ingredients from active packaging and development of dedicated methods: a study of two iron-based oxygen absorbers, Food Additives & Cont. 2003, 20(3), 291–299.

Regulation (EC) No. 178/2002 of the European Parliament and of the Council of 28 January 2002 laying down the principles and requirements of food law, establishing the European Food Safety Authority and laying down procedures in matters of food safety. Official Journal, L31, 1/2/2002, 1–24.

Regulation (EC) No. 1882/2003 of the European Parliament and of the Council of 29 September 2003 adapting to Council Decision 1999/468/EC the provisions relating to committees which assist the Commission in the exercise of its implementing powers laid down in instruments subject to the procedure referred to in Article 251 of the EC Treaty, Official Journal, L 284, 31/10/2003, P. 0001–0053.

Regulation (EC) No. 1935/2004 of the European Parliament and the Council of 27 October 2004 on materials and articles intended to come into contact with food and repealing Directives 80/590/EEC and 89/129/EEC. Official Journal of the European Union, L338, 13/11/2004, p. 4–17.

Rijk, R. and De Kruijf, N., Active and intelligent packaging development. Part I. Definitions, efficacy tests and migration experiments, Food, Cosmetics and Drug Packaging, 2003, 8, 152–159.

Verpakkingen-en gebruiksartikelenbesluit (Warenwet) Last updated Staatsblad 420, besluit van 30 mei 2005. Consolidated by Koninklijke Vermande, P.O. box 20014, 2500EA Den Haag, The Netherlands on December 2005.

Chapter 10

PACKAGING FOR NONTHERMAL
FOOD PROCESSING: FUTURE

Jung H. Han

Introduction

This book covers many topics on food packaging and nonthermal processes. Although the topics include innovative high technologies and advanced functions of food packaging and nonthermal processes, all of these packaging and processing technologies should satisfy the basic requirements of food packaging and processing. Food products are processed and packaged for the purpose of quality maintenance, product dispensing, and consumers' convenience. Since the basic requirements and the main purpose of packaging and processing food products should be satisfied, which yields the first priority, it is always important that the new technologies should play a role in protection and preservation prior to other extra advanced functions.

Basic and Advanced Functions of Food Processing

Food processing contains a variety of unit processes which include high-temperature treatments (blanching, cooking, pasteurizing, sterilizing, extrusion, evaporation, and distillation), low-temperature treatments (chilling, refrigerating, and freezing), raw material preparations (cleaning, sorting, and peeling), size reduction (cutting, grinding, homogenization, and emulsification), mixing, separation (centrifugation, filtration, and extraction), fermentation, irradiation, dehydration, and packaging. All of these processes are designed to produce and preserve consumable food products. For example, the basic functions of food packaging are containment, protection, and preservation of packaged food products. Other than these functions, which have the purpose of convenience, sales promotion, distribution, information providing, and marketing may be called as advanced functions perceiving superficial roles. Basic functions of food processing have technical aspects of food-quality objectives, while the superficial functions are mostly socioeconomic variables for food business and marketing.

Most innovative technologies are designed generally to achieve better performance of food processing. Many of them for food packaging are oriented to the superficial functions. However, they still have the requirements of perfect performance of basic functions of food processes. The examples of innovative technologies for basic performance improvement are high-barrier packaging materials, flash cooling/freezing techniques, precise prediction of microbial thermal resistance, thermal distribution and flow profile of particulate foods in aseptic processing, and so on. All of these performances are related to the improved basic functions of food processes.

Less Processed Foods

More processing usually results in lower quality but longer preservation period. For the extremely maximized safety of processed foods, every produce must be overcooked, hermetically packaged, irradiated, and stored in a deep freezer. To enhance the safety of food products, these processes trade-off the initial quality of foods. Besides some special cases, most consumers prefer fresh or fresh-like taste and flavor, which can be perceived as less processed and more natural. Minimally processed food may satisfy the basic functional requirements of processing and packaging and also the superficial socioeconomic goals simultaneously. Many researchers work to find the optimum process conditions of the minimum numbers of unit operations to produce more fresh-like but relatively safe products. This can be one of the driving forces of research and development of nonthermal processing. Since nonthermal processing systems use energy to pasteurize food products at a temperature far lower than normal pasteurization or room temperature, the quality attribute of nonthermally processed foods is different from that of conventionally heat-treated food products.

Trends in Consumers' Preference and Technologies

Demographic Changes

The world is aging. Most children of baby-boomers (i.e., persons born between 1946 and 1964) are currently beyond their teens (having been born between 1997 and 1994). The third generation of boomers will be dominant by the year 2020, and over a half of the world's population will be either over 60 or under 10 (USCB 2005). Due to this extreme bipolar age distribution, health and nutrition will likely be of critical interest to consumers (Sloan 2005). Health and nutritional benefits of processed foods would be the major issue of food science and industry. How can we prepare current food-processing technologies for this future? What would be the advantageous characteristics of nonthermal processing systems for the health and nutritional benefits compared to the conventional thermal processing? Less thermal degradation of nutrients and health-related ingredients would be the most attractive nature of nonthermal processing. Nutraceutical and fortified food industry may utilize nonthermal processing units more popularly.

More women will work in the future, and they will not have the time to prepare full meals at home. Besides the health benefit of foods, convenience of meal preparation and serving will also be an important factor of consumers' purchasing behavior (Sloan 2005). Ready-to-eat meals, partially prepared foods, microwaveable meals, easy open/close containers, and take-out foods will be more popular in the future to satisfy busy householders. Food processors and technologists may have many questions regarding this convenience-oriented trend. Are the nonthermal processes ready for this convenience trend? What kind of characteristics of nonthermal processes would be attractive for the improved convenience of food products in the future? These questions of convenience and also ingredient degradation should be investigated for the common utilization of nonthermal processing in the future.

Technology Development and Energy Resources

Fast changes and improvements are remarkable in the area of information technology, communication technology, biotechnology, and nanotechnology recently. This advancement will progress continuously in the near and far future. These new technologies will affect the future of food processing and packaging. The most dramatic and also feasible changes in food packaging, for example, would be any new applications related to food safety and intelligent packaging. The use of radio frequency identification for precision management of supply and distribution chains would be common in the near future. The use of nanocomposite materials for food packaging will be feasible soon. Food-processing specialists may be questioned: how much these cutting-edged high technologies will affect food products and food processing including nonthermal processes and packaging, which are considered as one of not-so-high technologies? Are the high technologies cost-effective for food processing?

Energy price is also one of significant factors which can direct the future food-processing systems and operations. High oil price may increase the cost of thermal processing operation. Nonthermal processes do not use heat resources for the inhibition of microorganisms. They use electric energy, pressure, or other types of energy for treatments. However, these energies have been generated mostly from petroleum energy. The high oil price will increase the cost of these nonthermal processing operations as well as thermal processing operations. Is irradiation, then, a solution to avoid the influence of high oil price despite its huge installation cost?

The commercial utilization of nonthermal processes will be affected by the advancement of other high technologies, energy cost, and other influences. The potential use and advantage of nonthermal processes also highly depend on the nature of nonthermal unit operation systems, food materials, and consumers' preference.

Nonthermal Processing of Foods

Active Packaging

Conventional packaging encloses food products for the protection of the foods from physical, chemical, and biological hazards. The packages have not been actively functional or did not increase the level of food quality. Packaging process is one of the unit operations at the last stage of food manufacturing for retarding quality degradation during distribution and providing convenience of handling and identity of the packaged products. There is no active extra function in this protective packaging system. However, active packaging systems continue their active functions after packaging during distribution. If the conventional packaging is a unit process to maintain the quality of enclosed foods, active packaging is a continuing process to modify the quality attribute and characteristics of packaged foods. Therefore, active packaging could improve the quality and safety of packaged food products, while conventional packaging retards the quality degradation. Furthermore, active packaging is a nonthermal processes because the process does not use heat. Among many applications of active packaging, the selected applications which can be called as nonthermal processes because of the nonthermal function of microbial inhibition are antimicrobial packaging, oxygen scavenging packaging, and carbon dioxide emitter. The use of these active packaging applications in combination with other nonthermal food processes would create more overall benefit and enhance the level of quality and safety of packaged foods.

Antimicrobial packaging is a packaging system requiring corporative knowledge of packaging materials, design architecture, and incorporated antimicrobial agents, which can kill or inhibit spoilage and pathogenic microorganisms during storage and distribution automatically (Han 2000, 2003). This antimicrobial function can be achieved by incorporating antimicrobial agents in the packaging system (headspace or packaging materials) and/or using antimicrobial polymers (Han 2000).

It is very important to evaluate the efficiency of the antimicrobial agents against target microorganisms before constructing commercial antimicrobial packaging system. Antimicrobial activity can be measured by various methods. However, the antimicrobial activity determined with culture media does not quantify its exact antimicrobial effectiveness with real food systems. Compared to the composition of culture media, foods have more complex ingredients and their natural microflora. When the antimicrobial packaging system is utilized commercially, the antimicrobial activity of the packaging system should be evaluated by the viable cell-counting test with the real foods. Other than the activity, the most important aspect of the use of antimicrobial agents is to satisfy all regulations of authorized governmental agencies or associations. The antimicrobial agents should be food ingredients, food-grade additives, or food contact substances (food contact materials in the EU). They must satisfy all of the related regulations in order to be utilized as constituents of the antimicrobial packaging system.

Foods are complex system with multiple ingredients representing tremendously various chemical and physical characteristics. These food characteristics and storage conditions dictate the balance of microbial world in the food and provide the optimal environments to only a small number of microbial species. Because of these typical microbial characteristics of foods, we could predict the microflora and assign specific target hazardous microbes for food products. Properly designed antimicrobial packaging system, therefore, could eliminate the risk of pathogenic disease and spoilage by inactivating the predictable growth of potential pathogenic and spoilage microorganisms under the specific conditions of food composition and storage/distribution systems.

The use of antimicrobial packaging system in combination with other nonthermal processes such as PEF, HPP, irradiation, modified atmosphere packaging (MAP), and antagonistic culture would be more beneficial to food preservation and safety enhancement. These combinations are good examples of hurdle technology. The system should be constructed after careful considerations of the synergism and adverse effects of the nonthermal process conditions, that is, pressure or irradiation dose, to the residual antimicrobial effectiveness of the packaging system.

Oxygen scavenging system and carbon dioxide-emitting system can also suppress the growth of aerobic microorganisms, such as fungi, aerobic bacteria, and also parasites. They can be used for the antagonistic culture to provide more favorable environments to lactic acid bacteria and other fermentative microorganisms. Therefore, they can play the role of nonthermal pasteurization. Of course, the biggest benefit of these two systems is the retardations of oxidation and respiration of packaged foods. The oxidative degradation of post-processed food ingredients can be decelerated under the reduced oxygen concentration. Oxygen scavenging packaging can protect the packaged foods from oxidation and extend their shelf life, as well as control the microbial growth. Carbon dioxide emitter can lower the respiration rate of fresh produce and also retard oxidation. Emitted carbon dioxide inside the package can be dissolved in water and reduce the pH of water. This acidification can inhibit various microorganisms.

The potential of selected active food packaging for nonthermally processed foods is so great that the packaging can contribute to the quality and safety enhancement of the food products which are processed nonthermally. Studies on the synergic effects of the use of active packaging as a supplemental nonthermal process and also any possible adverse effects should be conducted carefully for the practical commercialization of active packaging.

Bioactive Edible Coatings

Coatings are a particular form of films directly applied to the surface of foods. Removal of coating layers may be possible; however, coatings are typically not intended for disposal from the coated food. Therefore, coatings are regarded as a part of the food product. The use of edible films and coatings as carriers of bioactive substance has been suggested as a promising application of active food packaging (Cuq, Gontard, and Guilbert 1995; Han 2000, 2001). Edible coatings are produced from edible biopolymers and food-grade additives. Film-forming biopolymers can be proteins, polysaccharides (carbohydrates and gums), lipids, or their mixture (Gennadious, Hanna, and Kurth 1997). Plasticizers and other additives are mixed with the film-forming biopolymers to modify the physical properties or functionality of the coating layer. Edible coatings enhance the quality of food products, protecting them from physical damages, chemical migration, and biological deterioration (Kester and Fennema 1986). Most commonly, edible coatings function as barriers against oils, gases, or vapors and as effective carriers of active substances, such as antioxidants, antimicrobials, colors, and flavors.

Oxidation of food ingredients is one of the major causes of quality deterioration of food products. Particularly, high-fat foods or fried foods are very susceptible to lipid oxidation. Many functional food ingredients and nutraceuticals also contain highly oxidative chemical compounds. Such compounds include natural flavors, colorants, lipids, antioxidants, antimicrobials, and other bioactive chemicals. The oxidation of these compounds can be retarded by the addition of antioxidant chemicals through delaying onset or slowing the reaction kinetics of the oxidation (Nawar 1996).

Antimicrobial containing coatings have more advantages than the direct addition of the antimicrobial agents into foods because the coating can be designed to cause the slow release of antimicrobials from the coating layer. By slow diffusion of antimicrobials into food, the preservative activity of coating layer is maintained effectively above minimal inhibitory concentration at the surface of the foods. Thus, smaller amount of antimicrobials would be enough in an edible coating layer to achieve a target shelf life as compared with dipping, dusting, or spaying antimicrobial agents on the surface of food. Major potential food applications of antimicrobial coatings are to the perishable foods including fish, poultry, bakery goods, cheese, fruits, and vegetables.

It is important to carry out studies on the migration kinetics and control of the agents as much as the studies on the agents and their compatibility with carrier coating materials. The diffusion of the agents from coating materials into foods should be controlled slowly with respect to the kinetics of microbial growth. Slow release can prevent the growth of microorganisms even during the period after the package is opened or, more seriously, when it is defected. Since the antimicrobial agents are initially located in the coating layer of foods, which is a part of edible portion, the agents should be approved and also claimed as food ingredients. This is also one reason why many researchers are working with natural antimicrobial agents.

Edible films and coatings are promising systems for the improvement of food quality, shelf life, safety, and functionality. They can be used as individual packaging materials, food coating materials, active ingredient carriers, or separating compartments for heterogeneous ingredients within foods. Especially, the antimicrobial coating system which has shelf-sterilizing functions on the food surface can simplify sterilization processes for the minimally processed foods or fresh-like produce. Combination of antimicrobial edible coating with other nonthermal processing units will be a good hurdle technology.

Antagonistic Culture and MAP

MAP is a technology to control the internal gas composition of packages. Specific gas composition is obtained by the dynamic balance between respiration of foods (or gas consumption or production from packaged foods) and permeation of gases through packaging materials. Most successful commercialization has been achieved for the packaging of fresh or minimally processed produces and meat products. Due to the consumption of oxygen and production of carbon dioxide from respirating produce, the internal atmosphere of packages turns to reduced oxygen and elevated carbon dioxide level during storage. The change of gas composition is equilibrated at certain level, because the packaging materials are gas permeable. MAP cannot increase the quality of packaged foods; it can only reduce the respiration to extend the period over the threshold quality level of the products and maintain the products acceptable longer. MAP is not the tool to control the growth of microorganisms, although it has some activity to inhibit the growth of restrictively aerobic microorganisms such as fungi. Therefore, MAP is not classified as nonthermal processing applications.

MAP can be a very useful tool for nonthermal processing. Antagonistic culture of lactic acid bacteria in perishable foods under the MAP condition can provide favorable gas composition to the inoculated lactic acid bacteria or other harmless microorganisms. The growth of lactic acid bacteria can reduce the pH of the foods, deplete the essential nutrients, compete with pathogenic or spoilage bacteria, and, therefore, control the growth or viability of the undesirable microorganisms. Lactic acid bacteria, and others also, produce bacteriocins which are generally antimicrobial peptides including short-chain proteins or lipopeptides. They can inhibit the growth of other competing bacteria. Example bacteriocins of lactic acid bacteria are lacticin and pediocin. Other than bacteriocins, some bacteria produce antimicrobial organic acids such as propionic acid in Swiss cheese which inhibit the growth of spoilage microorganisms. Applicable examples of the use of antagonistic culture for the preservation of packaged foods are acetic acid and ethanol produced by inoculated bacteria and yeast.

The most important factor of MAP system is temperature maintenance. All MAP have been designed for the targeted storage temperature, mostly refrigeration temperature. Therefore, simple temperature abuse creates serious problem to the MAP because increased temperature accelerates respiration rate, chemical reactions related to quality degradation, moisture loss, microbial growth, and gas penetration through packaging materials. The change in temperature also alters the balance of antagonistic culture.

Lactic acid bacteria can survive at refrigeration temperature, but their growth under the refrigeration is negligible. Temperature abuse accelerates the growth of all microorganisms; however, if the microflora has been altered to be dominated by lactic acid bacteria after inoculation, the growth of lactic acid bacteria at abuse temperature is faster than other minor microorganisms. In the case of temperature abuse, the antagonistic culture in MAP shows more potential power to protect the packaged foods when the food has been inoculated with lactic acid bacteria.

MAP can be more beneficial to preserve the quality and safety of the packaged foods when they are applied in combination with other nonthermal processing. As an example, MAP with high-pressure processing (HPP) may have synergic effects on some produce. This combination, however, may have other unexpected adverse effects because the pressurization may change the respiration profile of packaged foods and also the permeation of gas through the packaging materials. MAP with irradiation also alters the respiration of produce and meats and also differs the gas permeability of packaging materials. As special considerations are required to design MAP differently from each commodity, in-depth studies on the effects of other nonthermal processes on the properties of packaged produce and meats and packaging materials should be investigated.

Packaging for Pulsed Electric Field (PEF) Processing

PEF process is generally designed with aseptic processing. Therefore, basic studies of aseptic processing and packaging should be accurately applied for the PEF-aseptic processing technology. For the aseptic packaging, high barrier is always required for preventing seal defects and pin hole/crack formation. These physical defects can provide the chance of microbial invasion and contamination. However, in the case of PEF-aseptic packaging, the invasion of oxygen also decreases the quality of PEF-treated food significantly. Therefore, PEF-aseptic processing requires very high-barrier packaging to prevent any physical defects and oxygen permeation. Most aseptic packaging materials are definitely high-barrier materials. The invasion of oxygen gas through the packaging materials under the normal situation is not a practical problem. The oxygen penetration could happen through the sealing area or an extremely extended part of thermoplastic molding system. Therefore, the end and side seals of aseptic carton are the most sensitive parts of oxygen penetration in the case of aseptic brick packages. For the case of aseptic bottle, the closure area and most thin part of the bottle are vulnerable to oxygen penetration. The prevention of oxidation for the packaging of PEF-treated foods is the key to the successful design of the process and packaging.

Some PEF-processed foods show accelerated oxidation kinetics compared to thermally processed foods. In order to understand this oxidation process and prevent the oxidation of PEF-processed foods, more serious analytical food chemistry studies are required.

Packaging for HPP

The selection of packaging materials for the HPP is very important than other nonthermal processes because foods have to be packaged before pressurization processing. Packaging materials should be recovered from pressurized deformation reversibly after the pressure is removed. Flexible or semirigid materials must be used for the pressurizing system, without any loss in their physical strength. The integrity of whole packages may be damaged at the sealing area or lamination. Seal defect or delamination would happen after HPP. More studies on the causes of seal defect and delamination by pressurization should be investigated, and the prevention protocol should be established.

No headspace or minimal headspace volume is required for the HPP packaging because the headspace causes deformation due to the compression of the gas. For the case of semirigid packaging such as bottles or containers, the complete headspace elimination would be more difficult compared to the flexible packaging. Also the hermetic sealing of the semirigid bottle cap could be defective under the pressure change during the process. Special design for securing the seal of caps would be required.

Oxygen permeability of packaging materials is a very important factor of oxidation of the foods. HPP should use high oxygen-barrier packaging materials. However, the oxygen permeability through the polymeric packaging materials increases dramatically after high-pressure treatment. It is recommended to select packaging materials which have oxygen permeability of 10 or 100 times lower than the calculated values of desirable materials. The reason to increase oxygen permeability and the magnitude of the increase after high-pressure treatment should be identified for each polymeric material.

Packaging for Irradiation

Foods are irradiated after packaging by gamma radiation, e-beam, or intense light. The irradiated beams kill microorganisms and change physiological responses of foods; however, they can also change the chemical structures of polymeric packaging materials. Irradiation of plastic materials can produce gases, low-molecular-weight hydrocarbons, halogenated polymers, and free radicals through the polymer scission process, crosslinking process between polymers, or radiolysis of additives. These polymer scission, crosslink reactions, and radiolysis alter the mechanical and physical properties of polymeric materials. Permeability of gases may be changed after irradiation. Free radical and peroxide, that is, radiolysis products, formation also causes quality degradation of packaging materials as well as foods. Other volatile hydrocarbons can be produced from polymers or additives after irradiation. Most polyolefins have stable chemical structures against 10-kGy irradiation. Vinyls are not recommended or permitted for the use of packaging materials for irradiated foods, since they have a serious polymer scission phenomenon after irradiation.

The effects of irradiation at various doses on the physical and chemical properties of commercial polymers have been investigated and quantified scientifically. The studies on the irradiation effects to the small amount plastic additives may be required whenever new additives are blended into plastics. The migration studies of radiolysis products into food should be completed before the commercial use of packaging materials in the case of food contact packaging materials. All new materials are required to be tested for their chemical migration and toxicology assessment after irradiation. To support the wide commercial use of irradiation for packaged foods, various regulations should also be established and updated. More detailed protocol for testinxg, filing, and application should be developed through collaborations with industry, regulatory agencies and scientists.

Conclusions

Future is always uncertain. Therefore, the diverged utilization of unit processes is recommended for better control of process, quality, and business management. Nonthermal processes may or may not substitute whole thermal processes. However, they will be used more frequently for food processing. The issue of packaging studies for the nonthermally processed foods is very important to preserve the quality of the extended shelf-life food products. These packaging studies may include the physical and chemical changes of packaging materials before/after nonthermal processes and the chemical interactions, such as migration, between nonthermally processed food ingredients and packaging materials. Regulations and consumer studies should provide the guidelines for the research and development of nonthermal processes and their packaging materials. Therefore, a safe implementation of nonthermal processes and their packaging for food products should be conducted through tight collaboration between scientists, industry, and regulatory agencies.

Convenience and safety would be the key topics for food and packaging industry in the near and far future. The research and development for nonthermal food processing and packaging should be directed to create the consumers' benefit regarding their convenience and safety enhancement.

References

Cuq, B., Gontard, N., and Guilbert, S. 1995. Edible films and coatings as active layers. In: *Active Food Packaging* (M. Rooney, ed.), pp. 111–142. Glasgow, UK: Blackie Academic & Professional.

Gennadious, A., Hanna, M.A., and Kurth, L.B. 1997. Application of edible coatings on meats, poultry and seafoods: a review. *Lebensm. Wiss. u. Technol.* 30(4): 337–350.

Han, J.H. 2000. Antimicrobial food packaging. *Food Technol.* 54(3): 56–65.

Han, J.H. 2001. Design edible and biodegradable films/coatings containing active ingredients. In: *Active Biopolymer Films and Coatings for Food and Biotechnological Uses* (Park H.J., Testin R.F., Chinnan M.S., and Park J.W., eds.), Proceedings of Pre-congress Short Course of IUFoST (April 21–22, 2001, Seoul, Korea), pp. 187–198. IUFoST.

Han, J.H. 2003. Antimicrobial food packaging. In: *Novel Food Packaging Techniques* (R. Ahvenainen, ed.), pp. 50–70. Cambridge, UK: Woodhead Publishing Ltd.

Kester, J.J., and Fennema, O.R. 1986. Edible films and coatings. In: *Food Proteins and Their Applications* (Damodaran S. and Paraf A., eds.). pp. 529–549. New York: Marcel Dekker.

Nawar, W.W. 1996. Lipids. In: *Food Chemistry* (Fennema., O.R., ed.), pp. 225–319. New York, NY: Marcel Dekker, Inc.

Sloan A.E. 2005. Demographic directions: Mixing up the market. *Food Technol.* 59(7): 34–45.

USCB. 2005. Data from US Census Bureau, Economic and Statistics Admin., US Department of Commerce, Washington, DC (http://www.census.gov).

INDEX

227